Oceans of Innovation

The Mathematical Heart of Complex Systems

Pravir Malik, Ph.D.
Professor Dr. Leon Pretorius

ISBN-13: 978-0-9903574-3-8

To Sri Aurobindo

Author's Note

We live in an infinitely innovative world. It is part of the very fabric of our existence. Being that this ability to innovate can express itself in massively different ways there are surely therefore a massive number of different approaches to express and frame innovation.

Here is one approach to frame such innovation. It has both an artistic and scientific dimension to it.

Artistic through necessity - in that when one tries to enter into the reality of what innovation is there is no expression that can fully represent it, and hence art seems at least to capture the current surge of such a contemplation. But the medium of the art here is mathematical since for me that alone seemed to capture the richness implicit in this de facto aspect of reality.

The mathematics expressed here is therefore the art. And this will be abundantly apparent in the attempt, for instance, to capture the reality of a point-instant infinite replete with all possibility. Necessarily the mathematical notation and the resulting mathematics had to expand beyond current norms in mathematics. This may raise eyebrows amongst some mathematicians. But as with my invitation in other recent writings I would welcome such individuals into collaboration to help capture perhaps in more robust ways what the artistic-mathematics here is seeking to express.

The mathematics expressed here is also a science. This is so because it proposes a structure and process to innovation that can be corroborated through observation and experiment. In its scientific aspect this book starts with a derivation of some subjective qualities of light that is proposed to set up objective reality. There is a four-foldness that is seen in this analysis that becomes the basis of the entire mathematics through the rest of the book. Hence the development of the mathematics primarily proceeds through induction, subsequently employing deduction to derive a representative set of mathematical operators that must be true of any complex system.

This book is therefore an artistic-mathematical contemplation of the aspect of innovation-reality that exists in every point-instant of any complex system. To make the most of it my suggestion is to suspend any judgment and enter into the unfolding contemplation to the degree possible. In this way you may see as I have seen and this may enrich in some way your own point of view.

Pravir Malik,
San Francisco

Chapter 1: The Light-Innovation Matrix

Innovation at a point, at any point in a system, will essentially be dependent on the nature of reality in that system. But the nature and essential possibility of reality is arguably determined by fundamental factors that precede human intervention. Here, therefore, we briefly examine the role that light may have played in setting up the fundamental landscape of reality, and based on that, what innovation could possibly mean.

So stepping back, it is perhaps fair to say that the speed of light has significant implications on the experienced nature of reality. The finiteness, c, at 186,000 miles per second in a vacuum (Perkowitz, 2011), creates an upper bound to the speed with which any object may travel also implying that objective reality will be experienced as a past, a present, a future from the point of view of that object (Einstein, 1995). These characteristics – a past, a present, a future – are implicit in the nature of light and become part of objective reality because of the speed of light. So the way we experience time seems definitely to be determined by c. This means too that the way innovation will be experienced, or will surface in time and through time, has also been in some sense determined by c.

Further, c also creates a lower bound when inverted (1/c) being proportional, or arguably even determining Planck's constant, h, that pegs the minimum amount of energy or quanta required for expression at the sub-atomic level. Planck's constant, h, pegging the amount of energy required for expression, therefore allows matter to form (Lorentz, 1925) and for the reality of nature with a past, present, and future, to also be experienced as a phenomena of connection between seemingly independent islands of matter. This characteristic of 'connection' is therefore also proposed to be implicit in the nature of light and becomes part of objective reality because of the speed of light. So in some sense the entire landscape of innovation has been bounded by c.

But a 'past' can also be viewed as established reality as defined by what the eye or other lenses of perception can see. It is the perceivable result of all the work and effort that has taken place so far. It is the foundation upon which the present and future will be built. It represents a status quo, a stability, and even a rigidity, and given that it is the result of the long play of time, it will not easily be persuaded to become another thing. It can be thought of as that which the eye can see when it looks around it. There is "physicality" to what the eye can see and so the essence of the past is a kind of physical-ness. So, ingrained in light, is this ability to project or create physical-ness.

The 'present' is the tremendous play of forces of all kinds to express themselves here and now. There is "vitality" that is present in this play and often it is the most energetic or forceful of the forces that will win out, as opposed to the most insightful or thoughtful. All the tremendous possibility of the future is seeking for expression now and so this essential vitality can also be thought of as a projection or possibility implicit in light.

The 'future' is the inevitability of what will manifest. The great thoughts, the great ideas, the purpose, the possibilities will sooner or later express themselves in what we call the future. And the essence of this is thoughtfulness or a curiosity or a purpose that we can summarize as an essential "mentality". So embedded in light is this ability to project mentality.

These implicit characteristics of the nature of light as experienced at the layer of reality so set up by a finite speed of light may hence be summarized by Equation (1), where c_u refers to the speed of light of 186,000 miles per second, that has created the perceived nature of reality, U, as expressed in Equation 1.1:

$$c_U: [Physical, Vital, Mental, Connection]$$

Eq. 1.1: Implicit Characteristics of the Nature of Light

So certainly some of the substance that innovation can play with is physical, vital, mental, and has to do with connection.

Exploring further, it is known however that at quantum levels the nature of reality is characterized by wave-particle duality. Light itself (Feynman, 1985) and matter (De Broglie, 1929) may be experienced as both particles and waves (Ekspong, 2014). But for matter to be experienced as waves implies that 'h' must have become a fraction of itself, $h_{fraction}$ to allow the concentration or possibility of quanta to have dispersed into wave-form. This further implies that c must have become greater than itself, c_N, such that the inequality specified by Equation (1.2) holds:

$$c_N > c_U$$

Eq. 1.2: Layer N, Layer U Inequality of Speed of Light

Note that what is implied here is that just as there is a nature of reality specified by U that is the result of the speed of light being 186,000 miles per second, so too there is another nature of reality specified by N that is the result of a speed of light greater than 186,000 miles per second.

This is akin to recent developments in physics with the notion of property spaces being separate from but influencing physical space as explored by Nobel Physicist Frank Wilczek (Wilczek, 2016). But further in "Slow Light" Perkowitz's recent treatment of today's breakthroughs in the science of light (Perkowitz, 2011) he states: "Although relativity implies that it's impossible to accelerate an object to the speed of light, the theory may not disallow particles already moving at speed c or greater. In the 1960's, Olexa-Myron P. Bilaniuk of Swarthmore College and E.C. George Sudarshan at Syracuse University began considering how to fit what they called "metaparticles" with speeds greater than c into the relativistic scheme. The approach was extended in 1967 by Gerald Feinberg (Feinberg, 1970), or Rockefeller and Columbia Universities, in his theoretical paper "Possibility of Faster-Than-Light-Particles," Feinberg also introduced the wonderful name "tachyons" for these hypothetical particles, from the Greek word "tachys" meaning swift." Perkowitz goes on to say how a flurry of papers have continued to appear about tachyons.

So light traveling at c_N seems to be possible. Current instrumentation, experience, and normal modes of thinking though, having developed as a bi-product of the characteristics so created in the layer of reality U may be inadequate to access N without appropriate modification. The notion of wave-particle duality already challenges the notion of normal thinking perhaps because wave-like phenomena could be a function of faster than c motion, and particle-like phenomena a function of less than or equal to c motion. That these may be happening simultaneously is reinforced by principles such as complementarity in which experimental observation may allow measurement of one or another but not of both (Whitaker, 2006).

But then taking this trend of a possible increase in the speed of light to its limit, this will result in a speed of light of infinite miles per second. The question is, what is the nature of reality when light is traveling at infinite miles per second? In any space-time continuum light originating at any point will instantaneously have arrived at every other point. Hence light will have a full and immediate *presence* in that space-time continuum. Further, that light will *know* everything that is happening in that space-time completely instantaneously – that is know what is emerging, what is changing, what is diminishing, what may be connected to what, and so on - or have a quality of knowledge. It will connect every object in that space-time completely and therefore have a quality of connection or *harmony*. Finally nothing will be able to resist it or set up a separate reality that excludes it and hence it will have a quality of *power*.

These implicit characteristics of the nature of light as experienced at the layer of reality so set up by an infinite speed of light may hence be summarized by Equation (1.3), where c_{∞} refers to the speed of light of ∞ miles per second, that has created the perceived nature of reality, ∞:

$$c_{\infty}: [Presence, Power, Knowledge, Harmony]$$

Eq. 1.3: Implicit Characteristics of Nature of Light at ∞ Speed

Equation (1.3) then seems also to provide insight into the playing field of innovation and will be explored further. It seems to suggest a limit or a starting point or an ending point for what is possible with innovation. But by (1.3) it can also be noticed that 'physical' is related to presence, 'vital' is related to power, 'mental' is related to knowledge, and 'connection' is related to harmony.

The question then, is how do these apparent qualities at ∞ precipitate or become the physical-vital-mental-connection based diversity experienced at U? This may be achieved through the intervention or action of a couple of mathematical transformations acting on the implicit characteristics of nature of light at ∞ speed as summarized by Equation (1.3). First, the essential characteristics of Presence, Power, Knowledge, Harmony that it is posited exist at every point-instant by virtue of the ubiquity of light at ∞ will need to be expressed as sets with up to infinite elements. Second, elements in these sets will need to combine together in potentially infinite ways to create a myriad of seeds or signatures that then become the source of the immense diversity experienced at U. This suggests that all that is seen and experienced at U may be nothing other than 'information' or 'content' of light and as such that there are fundamental mathematical symmetries at play where everything at U is essentially the same thing that exists at ∞.

Assuming that the first transformation occurs at a layer of reality K where the speed of light is c_K, such that $c_U < c_K < c_\infty$, this may be expressed by Equation (1.4):

$$c_K: [S_{Pr}, S_{Po}, S_K, S_H]$$

Eq. 1.4: The First Transformation at Layer K

S_{Pr} signifies 'Set of Presence', S_{Po} signifies 'Set of Power', S_K signifies 'Set of Knowledge', S_H signifies 'Set of Harmony/Nurturing'.

Assuming that the second transformation occurs at a layer of reality N where the speed of light is c_N, such that $c_U < c_N < c_K < c_\infty$, this may be expressed by Equation (1.5):

$$c_{N:}\, f(S_{Pr}\ x\ S_{Po}\ x\ S_K\ x\ S_H)$$

Eq. 1.5: The Second Transformation at Layer N

The unique seeds are therefore a function, f, of some unique combination of the elements in the four sets S_{Pr}, S_{Po}, S_K, S_H.

The relationship between the layers of light may be hypothesized by the following matrix in Equation (1.6):

$$Light_{Matrix} = \begin{vmatrix} c_\infty: [Pr, Po, K, H] \\ (\downarrow\ R_{C_K} = f(R_{C_\infty})) \\ c_K: [S_{Pr}, S_{Po}, S_K, S_H] \\ (\downarrow\ R_{C_N} = f(R_{C_K})) \\ c_N: f(S_{Pr}\ x\ S_{Po}\ x\ S_K\ x\ S_H) \\ (\downarrow\ R_{C_U} = f(R_{C_N})) \\ c_U: [P, V, M, C] \end{vmatrix}$$

Eq. 1.6: Light Matrix

The matrix should be read from the top row down to the bottom row as indicated by the ↓ between rows, and suggests a series of transformations leading from the ubiquitous nature of light implicit in a point – presence, power, knowledge, harmony - to the seeming diversity of matter observed at the layer of reality U which is fundamentally the same presence, power, knowledge, and harmony projected into another form of itself.

The first transformation is summarized by Equation (1.7):

$$R_{C_K} = f(R_{C_\infty})$$

Eq. 1.7: Light Matrix First Transformation

This is suggesting that the reality at the layer specified by the speed of light c_K, R_{C_K} is a function of the reality at the layer specified by the speed of light c_∞. This transformation translates the essential nature of a point into the sets described in Equation (1.4).

The second transformation is summarized by Equation (1.8):

$$R_{C_N} = f(R_{C_K})$$

Eq. 1.8: Light Matrix Second Transformation

This is suggesting that the reality at the layer specified by the speed of light c_N, R_{C_N} is a function of the reality at the layer specified by the speed of light c_K. This transformation combines elements of the sets into unique seeds as suggested by Equation (1.5).

The third transformation is summarized by Equation (1.9):

$$R_{C_U} = f(R_{C_N})$$

Eq. 1.9: Light Matrix Third Transformation

This is suggesting that the reality at the layer specified by the speed of light c_U, R_{C_U} is a function of the reality at the layer specified by the speed of light c_N. This transformation builds on the unique seeds suggested by Equation (1.5) to create the diversity of U as specified by Equation (1.1).

In this framework the notion of wave-particle duality hence may become complementary block-field-wave-particle "quadrality" where block refers to phenomenon resident to ∞, field to phenomenon resident to N, wave to phenomenon resident to K, and particle to phenomenon resident to U.

But an ocean is also a metaphor for block-field-wave-point quadrality. The block is all the reality always present behind the surface and is captured by

Equation (1.3) or the top line in the Light Matrix as expressed in Equation (1.6). The field is captured by Equation (1.4) or the second line from the top in Equation (1.6) and can be thought of as layers of possibility existing in each of the sets. The wave is captured by the creation of seeds represented by Equation (1.5) or by third line in Equation (1.6). The particle that is apparently disconnected from the whole is captured by Equation (1.1) or the bottom line in Equation (1.6).

So what we arrive at is a fundamental light-innovation matrix that essentially sets up the context and meaning of innovation. Let us further explore this now.

Chapter 2: Implicit Order in Time and Space

The basic four-foldness derived at in Chapter 1 through a consideration of the speeds of light can be arrived at through quite a different process as well.

In Connecting Inner Power with Global Change: The Fractal Ladder (Malik, 2009), an exploration into fractal patterns and biases at levels of increasing system complexity, the case is made that there is a similar pattern that animates progress regardless of the type of organization. An organization, hence, could be an individual, a team, a business unit, a corporation, a market, a country, planet earth, or a system of thought, for example. This pattern is summarized as the movement from the 'physical' to the 'vital' to the 'mental', where the physical can be thought of as reality as characterized by what the eye can see. Necessarily the eye sees what has already been created: hence the physical can also be thought of as an orientation reinforcing the status quo. The vital can be thought of as reality as characterized by the play of energy. Such play often results in the strongest energy winning out, and hence the vital can be thought of as an orientation reinforcing experimentation even bordering on aggression. The mental can be thought of as reality as experienced by thought. Hence this orientation can be thought of as encompassing curiosity, idealism, and the future. In How to Create a Mind (Kurzweil, 2013) Kurzweil maps parts of the brain to similar functions. Hence the mental function is primarily fulfilled by the neocortex, the vital function primarily by the amygdala and hippocampus, and the physical function primarily by the cerebellum.

An organization can center itself in any of the three phases, and will perceive the world, interpret circumstances, and act consistently with the psychology that emanates from that orientation. As an organization progresses though, to thereby increase its degrees of freedom, it is observed that it essentially changes its orientation from the physical to the vital to the mental (Malik, 2009). Further, it is possible to bring about an organization's transformation by causing the shift from the physical, to the vital, to the mental (Malik, 2003). In his book, Exponential Organizations, Singularity University's Ismail also suggests this trend from the 'physical' to an 'information-orientation' (Ismail, 2014) and how the information-orientation allows an organization to be "ten times better, faster, and cheaper than yours". The surfacing of such a trend from the physical to the vital to the mental, regardless of area suggests that any CAS may have this underlying direction.

If this is so, then the notions of 'chance' and 'probability' may also be enhanced. It is possible that things are not random, but rather are the outcome of a working out of a number of forces vying for expression. If so, chance and probability may be thought of as areas where meta-levels, as in the levels after the bottom-level in (1.6) act. This will be further explored in Chapter 11, Qualified Determinism in Complex Adaptive Systems.

This change in orientation from the physical, to the vital, to the mental is perceived as being the driver of innovation, and the hypothesis is that the orientations – the physical the vital, the mental – are embedded in the warp and woof of any CAS. Just as the bases of computer-technology is a binary representation and manipulation of zeros and ones, the hypothesis for a broad innovation-technology is a trinary representation and manipulation of the physical, the vital, and the mental. This is so because the physical, the vital, and the mental concisely summarize the orientation of a system and therefore the possibilities available to it.

Many examples of a change in orientation leading to an opening of possibility have been recorded time and time again. Referring to the author's previous work – Business Transformation through the Creation of a Complex Adaptive System (Malik, 2003), An Introduction to Fractal Dynamics (Malik, 2004), Connecting Inner Power with Global Change (Malik, 2009), An Integral Perspective on Current Economic Challenges: Making Sense of Market Crises (Malik, 2013), The Flower Chronicles: A Radical Approach to System and Organizational Development (Malik, 2014) and The Fractal Organization (Malik, 2015), as an indication of a global trend now being captured by more and more people in their respective fields. For example, as pointed out earlier Ismail traces the success of new tech-based companies such as Google, Facebook, Uber, Ebay, amongst others and attributes it to a shift from the 'physical' to an 'information' orientation (Ismail, 2014). In University of Pennsylvania's Seligman's observations (Seligman, 2011) there are three tiers that can determine happiness. The first is what he calls 'the pleasant life', the second 'the good or engaged life', the third 'the meaningful life'. "The pleasant life is about using savoriness and mindfulness to enhance the pleasures one experiences" (a physical-orientation). "The good life is about using strengths to increase flow to a point where time stops" (a vital-orientation). "The meaningful life is about using strengths in the pursuit of a goal far vaster than oneself "(a mental-orientation). In thousands of observations Seligman found that increasing pleasures has almost no impact on increasing happiness. On the contrary increasing meaning had the most impact, and increasing flow had a strong impact. It is only when meaning and flow both exist that increasing pleasure may matter. This finding is another instance of what is proposed as a universal, underlying trend in any system, of the shift by its actors from a physical to a vital to a mental orientation in order to increase respective degrees of freedom. In Donella Meadow's observations of most effective places to intervene in a system (Meadows, 2008) she identifies twelve levers and places them in order of their effectiveness.

Figure 2.1 summarizes these levers in increasing order of importance, and further identifies the nature of the lever in terms of the physical, vital, or mental orientation:

Increasing Order of Importance	Lever Name	Lever Description	Orientation
1	Numbers	Constants and parameters such as subsidies, taxes, standards	Physical
2	Buffers	The sizes of stabilizing stocks relative to their flows	Physical
3	Stock-and-Flow Structures	Physical systems and their nodes of intersection	Physical
4	Delays	The lengths of time relative to the rates of system change	Vital
5	Balancing Feedback Loops	The strength of the feedbacks relative to the impacts they are trying to correct	Vital
6	Reinforcing Feedback Loops	The strength of the gain of driving loops	Vital
7	Information Flows	The structure of who does and does not have access to information	Vital
8	Rules	Incentives, punishments, constraints	Mental
9	Self-Organization	The power to add, change, or evolve system structure	Mental
10	Goals	The purpose of the system	Mental

11	Paradigms	The mind-set out of which the system – its goals, structure, rules, delays, paradigms - arises	Mental
12	Transcending Paradigms	Strategically, profoundly, madly, letting go and dancing with the system	Mental

Figure 2.1 System Intervention Levers in Order of Importance and Relationship to Orientation [adapted from 'Thinking in Systems' (Meadows, 2008)]

It can be observed that as the effectiveness of a lever increases so does the underlying orientation shift from the essentially physical, to the vital, to the mental. In an article in Harvard Business Review describing how the Chinese internet company Alibaba (Reeves et al, 2015), uses algorithmic thinking to constantly reinvent itself, the crux of the algorithm is also a movement through three sequential phases: discovering what works (physical), adjusting how and how much experimentation takes place (vital), and influencing customer's preferences (mental).

In "At Home in the Universe" Stuart Kaufmann (Kaufmann, 1995) suggests that a combination of natural selection and self-organization leads to matter organizing itself into complex structures in spite of the forces of entropy. He discusses the second law of thermodynamics that has as a consequence the disappearance of order from equilibrium states. While entropy leads to a sense that an incoherent collapse of order is the natural state of things, he cites abundant evidence of order in our world, from microscopic cells, to plenitude of species, to the postmodern technological era with its exploding rate of innovation.

Coupled with the proposed system-states of the physical, the vital, and the mental, and the observation that it is only in the vital-type conditions that a trend of order can reverse itself, this may lead to the insight that the Second Law of Thermodynamics is related to vital-type conditions.

Further, he raises the question as to whether there are any general laws that govern this range of complex activity and further if this would allow us to achieve predictive powers. He cites two commonly accepted characteristics of CAS – indeterminism, and sensitivity of CAS to changes in initial conditions – to discard the possibility of achieving predictive powers. However, it is possible that some form or predictability could exist if some form of determinism existed.

Kaufmann's hope is to characterize classes of properties of systems that are generic and do not depend on the details. He suggests that it is possible to form theories of system behavior that are insensitive to a full set of descriptive details.

In stepping back and asking why this similar pattern of progress exists ubiquitously, what this may point to is a reality of an implicit order that occurs in Time and Space:

- Observation #1: This progress happens regardless of area of the world, and regardless of industry. This points to a characteristic of an implicit or system-presence, whereby wherever there is the possibility of progress it will happen.
- Observation #2: This progress happens in spite of tremendous opposition to it, again regardless of field and area. This points to a characteristic of implicit or system-power.
- Observation #3: The right instrumentation and the right circumstances seem to be leveraged in order that the progress possible does happen.

This points to a characteristic of implicit or system-knowledge that knows what to leverage and when. System-knowledge implies prerequisite-state that contains a synthesis of contraries, and a consequent-state such that the right component is invoked at the right time. The notion of synthesis of contraries is pressing to the surface in many fields of life. Grand Unified Theories are being worked on in mathematics and physics, for example. In mathematics Frenkel, in his book Love and Math (Frenkel, 2013), refers to the Langlands program as "a Grand Unified Theory of Mathematics because it uncovers and brings into focus mysterious patterns shared by different areas of math and thus points to deep, unexpected connections between them." In his book, Dreams of a Final Theory (Weinberg, 1994), written in the wake of US Congress deliberations and rejection of the Superconducting Super Collider, Nobel Laureate Weinberg suggests progress towards "a theory of everything (ToE)...a hypothetical single, all-encompassing, coherent theoretical framework of physics that fully explains and links together all physical aspects of the universe".

- Observation #4: The fact that this progress happens so that the degrees-of-freedom of the associated actors is continually increased points to an implicit or system-nurturing or system-harmony.

Hence, one can say that implicit in Space, there are four overarching characteristics – that of system-presence, system-power, system-knowledge, and system-harmony/system-nurturing (Malik, 2009). If then there is an implicit order in Time and an implicit order in Space, then the notion of organizational causality needs to be re-examined, since the very matrix in which any organizational play can arise is bounded differently than it may have so far been imagined or conceptualized.

It is also interesting to note that mainstream CAS thinkers are beginning to leverage space and time in more comprehensive ways in their interpretation of CAS. For example in her article, Life's Information Hierarchy (Flack, 2014), Flack suggests that the explanation for complex, multi-scale structure of biological and social systems lies in their manipulation of space and time to reduce uncertainty about the future. She states: "To many, hierarchical organizations suggests the nesting of components or individuals into groups, with these groups aggregating into yet larger groups. But this view – at least superficially – privileges space and matter over time and information. Many types of neural coding, for example, require averaging or summing over neural firing rates. The neurons' spatial location – that they are in proximity – is, of course, important, but at least as important to the encoding is their behavior in time."

Chapter 3: Nature of a Point-Instant in a System

The 'point-instant' captures the inherent innovation that appears to exist in the system and is represented by the top-line in Equation (1.6), the Light-Matrix. This innovation is a function of the properties of light derived in Equation (1,3) - presence, power, knowledge, and harmony – and reproduced here for convenience:

$c_{\infty}: [Presence, Power, Knowledge, Harmony]$

The 'point' aspect of the point-instant suggests the space-dimension and that space is seeded with the possibilities inherent in the properties of presence, power, knowledge, and harmony. The 'instant' aspect suggests the time-dimension and gives insight into the process of emergence that the possibilities in space progressively surface as.

In its point-instant, presence-power-knowledge-harmony wholeness, presence allows emergences to continue to develop as per the possibilities implicit in the past-present-future or physical-vital-mental pathway.

Beginning to translate this into an equation, the notation $System_{Pr}$ is given to system-presence. This system-presence is true across any considered Time-Space continuum starting from a time-space boundary '0' to a time-space boundary 'N'. This notion is characterized by the notation $TS_{0\rightarrow N}$. Within that boundary from 0 to 'N', the 'presence' is such that it will always seize an opportunity to cause a shift from the physical-leading to the vital-leading, and from the vital-leading to the mental-leading.

The notion that the presence seizes on opportunity as

characterized by the notation:

$$\begin{array}{c} Presence \\ \downarrow \\ Opportunity \end{array}$$

The shift from physical-leading to vital-leading and vital-leading to mental-leading is characterized by:

$$\begin{array}{ccc} P_L & \rightarrow & V_L \\ V_L & \rightarrow & M_L \end{array}$$

Hence in this approach it is suggested that:

$$System_{Pr} \equiv TS_{0 \rightarrow N} \left[\begin{array}{c} Presence \\ \downarrow \\ Opportunity \end{array} \begin{bmatrix} P_L & \rightarrow & V_L \\ V_L & \rightarrow & M_L \end{bmatrix} \right]$$

But there is something else about this presence as well. All other developments take place in it. That is, it provides a container of sorts in which the plays of system-power, system-knowledge, and system-harmony/system-nurturing can take place. This notion is summarized by the notation:

$$Container_{\begin{bmatrix} System_P \\ System_K \\ System_N \end{bmatrix}}$$

Hence, combining these various components, an equation for 'system-presence', Equation 3.1, arises:

$$System_{Pr} \equiv TS_{0 \rightarrow N} \left[\begin{array}{c} Presence \\ \downarrow \\ Opportunity \end{array} \begin{bmatrix} P_L & \rightarrow & V_L \\ V_L & \rightarrow & M_L \end{bmatrix} \ \& \ Container_{\begin{bmatrix} System_P \\ System_K \\ System_N \end{bmatrix}} \right]$$

Eq 3.1: System Presence

In its point-instant, presence-power-knowledge-harmony wholeness, power allows emergences to continue to happen in spite of tremendous oppositions of all kinds; this too, regardless of field or area.

Constructing an equation for system-power, the notation $System_P$ is used to represent system-power. Any endeavor will always be met with resistances of various kinds. The resistances that arise along the physical dimension are referred to as P_R. The resistances that arise along the vital dimension are referred to as V_R. The resistances that arise along the mental dimension are referred to as M_R. In the fruition of any endeavor one or all of these types of resistances may arise. Further, resistance of one kind often feeds on resistance of another kind, and to generalize the resistances encountered in an endeavor may be characterized as the product of the three types of resistance

$$P_R * V_R * M_R$$

These resistances arise across any considered Time-Space boundary from 0 to 'N', and therefore it may be said that the power of the system is such that:

$$power > \sum_{TS=0}^{N} P_R * V_R * M_R$$

An equation for 'system-power', Equation 3.2, hence, is the following:

$$System_P \equiv power > \sum_{TS=0}^{N} P_R * V_R * M_R$$

Eq 3.2: System Power

In its point-instant, presence-power-knowledge-harmony wholeness, knowledge orchestrates emergences to continue to happen by leveraging the right instruments and circumstances.

Translating this into an equation, the notation, $System_K$, is used for system-knowledge. This $System_K$ is such that it leverages the right instrumentation and circumstance to bring about the progress that is possible. This concept of 'instrumentation' is denoted by the subscript 'I'. The concept of 'circumstance' is denoted by the subscript 'C'. Both instrumentation and circumstance can be of a physical, vital, or mental type and this possibility is denoted by:

$$\begin{bmatrix} P_{I,C} \\ V_{I,C} \\ M_{I,C} \end{bmatrix}$$

Further, the notion that the 'knowledge' is such that it 'leverages' the right instrumentation and circumstance is depicted by:

$$\begin{matrix} Knowledge \\ \downarrow \\ Leverage \end{matrix}$$

This act of leveraging results in a fundamental shift so that the physical-leading yields to the vital-leading, and the vital-leading yields to the mental-leading. Hence:

$$\begin{matrix} Knowledge \\ \downarrow \\ Leverage \end{matrix} \begin{bmatrix} P_{I,C} \\ V_{I,C} \\ M_{I,C} \end{bmatrix} \rightarrow \begin{bmatrix} P_L & \rightarrow & V_L \\ V_L & \rightarrow & M_L \end{bmatrix}$$

Since this behavior may exist across any Time-Space

continuum an equation for system-knowledge, Equation 3.3, is suggested:

$$System_K \equiv TS_{0\rightarrow N}\begin{bmatrix} Knowledge & \begin{bmatrix} P_{I,C} \\ V_{I,C} \\ M_{I,C} \end{bmatrix} & \rightarrow & \begin{bmatrix} P_L & \rightarrow & V_L \\ V_L & \rightarrow & M_L \end{bmatrix} \\ \downarrow & & & \\ Leverage & & & \end{bmatrix}$$

Eq 3.3: System Knowledge

In its point-instant, presence-power-knowledge-harmony wholeness, harmony or nurturing allows emergences to continue to happen with more and more degrees of freedom to come to the surface.

This characteristic of this implicit-nurturing may be referred to as 'system-nurturing'. Like the other characteristics it is suggested to exist across a Time-Space continuum. This is depicted by:

$$TS_{0\rightarrow N}$$

There is an action of nurturing such that any state is always advanced to a higher level. This is depicted by:

$$\coprod_{Nurturing}\begin{pmatrix} P_- & M_+ \\ V_- & V_+ \\ M_- & P_+ \end{pmatrix}$$

Hence, there is a 'union', depicted by 'U' that 'nurtures' the negatives towards their positives.

Further, there is an increasing action of nurturing such that the possibility of integration is always increased to form a larger and larger basis. This increasing basis is depicted as being modulated by the polar coordinates 'r' and 'θ', where r is the radius which increases from an initial value of '0', and 'θ' is an angle from '0' to '360'.

This notion of an increasing of 'r' and 'θ' is reinforced by the relatively recent phenomena of 'Swift Trust' as a form of trust occurring in temporary organizational structures, which can include quick starting groups or teams.

Hence, the equation of system-nurturing, Equation 3.4, is depicted as:

$$System_N \equiv TS_{0\to N}\left(\coprod_{Nurturing}\begin{pmatrix} P_- & M_+ \\ V_- & V_+ \\ M_- & P_+ \end{pmatrix} mod\ (r,\theta)\right)$$

Eq 3.4: System Nurturing

It is suggested that these four characteristics exist across any system, and to denote this it is generalized that every point in any system is embedded with this four-fold intelligence. It is suggested that this four-fold intelligence is resident in every instant-spot of the system. It is suggested that to be able to leverage or activate this four-fold intelligence at will is the ultimate act of innovation.

Such a depiction of models, as Bar-Yam states in "From Big Data to Important Information" (Bar-Yam, 2016), 'is "valid" only because of the irrelevance of details...If we want to say anything meaningful about a system—meaningful in the sense of scientific replicability or in terms of utility of knowledge—the only description that is important is one that has universality, that is, is independent of details. There is no utility to information that is only true in a particular instance. Thus, all of scientific inquiry should be understood as an inquiry into universality—the determination of the degree to which information is general or specific.'

Chapter 4: Inherent System Bias

At a very basic level keeping in mind the three states – physical, vital, mental - six equations that provide approximations to situations such as Stagnation, Stability, Entropy, Energy, Sustainability, and Fragmentation are first suggested. These situations are particularly relevant to the management of technology and even sociotechnical systems that often comprises multiple directions and trends, such as stagnation, or stability, or fragmentation of technologies, for example, in any given segment. These equations are derived from the possibility and the obstacle provided by the three ubiquitous states.

Hence, if the physical orientation exists in its capacity as an obstacle, then it may be suggested that while the physical is leading (represented by the subscript 'L' for leading), it is doing so as an obstacle (represented by the superscript, '-') and hence the status quo will always remain as it is, and this will result in stagnation. This can be represented by the function in Equation 4.1:

$$Stagnation = fn\ (P_L^-)$$

Eq 4.1: Stagnation

On the flip-side, if the physical is leading in a progressive capacity this results in stability of established structure, and is a great foundation for other things to be built. Stability can be represented as in Equation 4.2:

$$Stability = fn\ (P_L^+)$$

Eq 4.2: Stability

If the vital is leading in its rapacious capacity, this will result in entropy as reflected in Equation 4.3:

$Entropy = fn\ (V_L^-)$

Eq 4.3: Entropy

In other words, with the random play of energy as captured by the Second law of Thermodynamics, the sum of the entropies of the participating thermodynamic systems increases (Van Wylen & Sonntag, 1985). Rather than existing as a universal state, the Second Law of Thermodynamics is true where the negative-vital is leading.

If the vital is leading in a balanced capacity, this will result in energy where and when it is needed as represented by Equation 4.4:

$Energy = fn\ (V_L^+)$

Eq 4.4: Energy

If the mental is leading so that an idea is pitted against another idea, then fragmentation will result. This is reflected in Equation 4.5:

$Fragmentation = fn\ (M_L^-)$

Eq 4.5: Fragmentation

If the mental is leading so that ideas are combined around the greatest idea, this will result in sustainability, as reflected in Equation 4.6:

$Sustainability = fn\ (M_L^+)$

Eq 4.6: Sustainability

Given these it can be suggested that the outcome of any circumstance is going to be a function of which one of these six states is the strongest, as represented by Equation 4.7:

$$Outcome_{circumstance} = Strongest\ (P_L^-, P_L^+, V_L^-, V_L^+, M_L^-, M_L^+)$$

Eq 4.7: Outcome of Circumstance

In general it is conditions of stability, energy, and sustainability as opposed to stagnation, entropy, and fragmentation, that are observed and hence it can be suggested that any system prefers these states and reinforces them.

Chapter 5: Architectural Forces

The characteristics embedded in a point as in (1.3) suggest a level of innovation that is hard to fathom. One can only glimpse the extraordinary nature embedded in a point. And yet it can be suggested that this extraordinary nature is barely visible unless the right analytical lens of the sort being suggested here is first set up. Further, it is suggested that this extraordinary nature is responsible for a broader set of architectural forces that exist behind the visible face of things.

Hence, system-presence, system-power, system-knowledge, and system-nurturing that define the nature of every point in our system, become more tangible as a broader set of architectural forces that emanate from each of them.

Considering system-presence, here is a characteristic that appears to be everywhere (Malik, 2009) at the service of all the constructs that develop within it. There is a diligence and perseverance by which any opportunity for progress is seized. Further, if one considers the extraordinary detail that appears in any construct, whether an atom, a body, a planet, or a galaxy, one is struck by the high degree of perfection that surfaces in this presence.

So if one contemplates the nature of this system-presence there is a set of forces that surface. Depicting such a set as $S_{System_{Pr}}$, one can arrive at elements such as Service, Perfection, Diligence, Perseverance, amongst others, that are part of this set. Hence, the set can be described by Equation 5.1:

$$S_{System_{Pr}} \ni [Service, Perfection, Diligence, Perseverance, \ldots]$$

Eq 5.1: Set of System Presence

Similarly, considering the characteristic of system-power, one can hypothesize that there is a family of forces that emanates from it. The kinds of forces may be thought of as Power, Courage, Adventure, Justice, amongst others. The set for system-power can hence be depicted by Equation 5.2:

$$S_{System_P} \ni [Power, Courage, Adventure, Justice, \dots]$$

Eq 5.2: Set of System Power

Similarly, considering the system-knowledge as the root of various powers that emanate from it, one may characterize the set for system-knowledge by Equation 5.3:

$$S_{System_K} \ni [Knowledge, Wisdom, Law\ Making, Spread\ of\ Knowledge \dots]$$

Eq 5.3: Set of System Knowledge

The set for system-nurturing is depicted by Equation 5.4:

$$S_{System_N} \ni [Love, Compassion, Harmony, Relationship\ \dots]$$

Eq 5.4: Set of System Nurturing

Chapter 6 describes how unique signatures or seeds for any type of organization can be built by leveraging the elements in the sets of architectural forces.

Chapter 6: Uniqueness of Organizations

The hypothesis is that every organization, whether an atom, cell, person, team, corporation, market, or country is unique and that this uniqueness can be specified in terms of elements of the derived sets for power, knowledge, presence, and nurturing. This hypothesis for uniqueness stems from observations at multiple levels.

At the sub-atomic level Nobel Laureate Wolfgang Pauli's 'Pauli Exclusion Principle' states that no two similar fermions, which include fundamental particles with half-integer spin such as protons, neutrons, and electrons, can occupy the same quantum states simultaneously (Pauli, 1964). Spin has to do with the angle that the particle has to rotate through before being symmetrical with its original state. Half-integer spin particles need to rotate through 720 degrees before being symmetrical with their original state. The implication of the Pauli Exclusion Principle is that fundamental structure and consequently stability comes into being at the atomic level, which as is evident in the periodic table allows the separation of function related to form. This stability related to the underlying structure of atoms implies the basis of uniqueness and diversity. In the absence of the Exclusion Principle matter would just be a dense soup (Hawking, 1988) with particles occupying overlapping space.

At the observable level uniqueness is evident from the immense diversity of distinct species on earth (Mora, 2011) estimated to be over 2 million, and further the uniqueness of every member of each species. This member-level uniqueness is suggested by the difference in non-coding regions of the DNA that may vary in their sequence by about 1 to 4 percent, which in turn result in unique protein binding sequences of each human

(Snyder, 2010), as an example, which in turn results in unique observable qualities.

At the astronomical level Einstein's Special Theory of Relativity (Einstein, 1995) suggests that every coordinate system potentially has its own space-time rendering as opposed to there being one absolute space and time. This implies the notion of uniqueness as an implicit property of space.

The four properties explored in the Section 4.2 define the source of that innovation. From this source emanate 4 sets of forces that suggest the boundaries of that innovation. It is also observed that no two organizations, regardless of scale, are alike (refer to Section 1.4). This uniqueness can be thought of as an additional mechanism of any system to ensure innovation.

Assuming then that the fount of uniqueness is system-presence, a general equation for organizations that belong to the family of system-presence can be derived. Such uniqueness can be depicted as Sig_x where the subscript 'x' refers to the source family, and 'Sig' or signature to 'uniqueness'. Hence the uniqueness of an organization in the family of system-presence would be notated by $Sig_{System_{Pr}}$.

In line with the development of properties of a point and the precipitating architectural forces as discussed in Chapter 2 and 4 respectively, an approach to constructing such uniqueness is to assume a primary factor X that drives the uniqueness that belongs to the set $S_{System_{Pr}}$. Further, assume that the uniqueness is qualified by a number of secondary factors Y that may belong to any of the 4 sets - $S_{System_{Pr}}, S_{System_P}, S_{System_K}, S_{System_N}$. The primary factor X would have a greater weightage than any of the secondary factors Y. The weightage of X hence could be depicted by the number 'a', and the weightage of Y a number $'b_{o-n}'$, such that a > b. Further, the secondary element can repeat from '0 – n' times, and is hence depicted as $\overline{Yb_{0-n}}$.

The equation, Equation 6.1, hence for a unique organization derived from the family of system-presence is:

$$Sig_P = Xa + \overline{Yb_{0-n}} \quad where \begin{bmatrix} X \in [S_{System_{Pr}}] \\ Y \in [S_{System_{Pr}}, S_{System_P}, S_{System_K}, S_{System_N}] \\ a, b \ are \ integers; a > b \end{bmatrix}$$

Eq 6.1: System Presence Based Unique Organization

Similarly, an equation, Equation 6.2, for a unique organization derived from the family of system-power is:

$$Sig_V = Xa + \overline{Yb_{0-n}} \quad where \begin{bmatrix} X \in [S_{System_P}] \\ Y \in [S_{System_{Pr}}, S_{System_P}, S_{System_K}, S_{System_N}] \\ a, b \ are \ integers; a > b \end{bmatrix}$$

Eq 6.2: System Power Based Unique Organization

An equation, Equation 6.3, for a unique organization derived from the family of system-knowledge is:

$$Sig_M = Xa +$$

$$\overline{Yb_{0-n}} \quad where \begin{bmatrix} X \in [S_{System_K}] \\ Y \in [S_{System_{Pr}}, S_{System_P}, S_{System_K}, S_{System_N}] \\ a, b \; are \; integers; a > b \end{bmatrix}$$

Eq 6.3: System Knowledge Based Unique Organization

An equation, Equation 6.4, for a unique organization derived from the family of system-nurturing is:

$$Sig_I = Xa + \overline{Yb_{0-n}} \quad where \begin{bmatrix} X \in [S_{System_N}] \\ Y \in [S_{System_{Pr}}, S_{System_P}, S_{System_K}, S_{System_N}] \\ a, b \; are \; integers; a > b \end{bmatrix}$$

Eq 6.4: System Nurturing Based Unique Organization

The four preceding equations can be generalized by Equation 6.5:

$$Sig = Xa + \overline{Yb_{0-n}} \quad where \begin{bmatrix} X \in [S_{System_{Pr}}, S_{System_P}, S_{System_K}, S_{System_N}] \\ Y \in [S_{System_{Pr}}, S_{System_P}, S_{System_K}, S_{System_N}] \\ a, b \; are \; integers; a > b \end{bmatrix}$$

Eq 6.5: Generalized Equation for Unique Organization

Having considered the structure of uniqueness, the next question is how does such uniqueness emerge? This is discussed in Chapter 7.

Chapter 7: Emergence of Uniqueness

While the uniqueness of organizations as represented by the Signature is a seed, like any seed there is a process for its emergence (Kaufmann, 1995; Portugali, 2012; Yates, 2012), and the uniqueness will often be hidden or very much behind the scene until certain conditions are fulfilled (Malik, 2009).

The implicit nature of Time and Space suggest a universal developmental model that provides a cue as to the process for emergence (Deep Order Mathematics Videos, 2016). In this model the four sets of architectural forces already described form a pool in space, as it were, from which possibility arises. Possibility itself is unique from point to point and is governed by the Equation for Uniqueness (Equation 6.1 through 6.5) described in the previous chapter.

Hence it is observed that initially the uniqueness takes a 'physical' form, moving on to a 'vital' form, and then onto a 'mental' form. Once the orientations implicit in each of these phases are assimilated, then the uniqueness takes on an 'integral' form. The integral form is a threshold phase, and allows the uniqueness suggested by the Signature to emerge in fuller force or in its 'force' form. The final phase is the 'contextual form' that allows the signature to act with impunity within a considered context.

Mathematically, if an organization exists at the physical phase, it may be suggested that its signature or uniqueness is modulated by the constant 'π'. π is the seed of a circle or sphere and can be thought of as defining behavior that is tightly bound. Within such a tightly bound volume it will likely not even be apparent what the uniqueness of an organization necessarily is. Assuming the uniqueness to be defined by the derived question *Sig*, the physical-level (P) behavior can be described by the following equation-segment where 'mod' signifies modulated-by:

$$P: Sig * mod\ (\pi)$$

If an organization exists at the vital level, it may be suggested that its uniqueness is modulated by the Euler-constant 'e'. e is at the root of exponential behavior. The vital by definition is about assertive and aggressive growth the symbol of which is 'e'. Hence vital-level (V) modulation (represented by 'mod') can be described by the following equation-segment:

$$V: Sig * mod\ (e)$$

If an organization exists at the mental level, it may be suggested that its uniqueness is modulated by the Gaussian Distribution 'G'. G summarizes rational behavior with a key direction followed by most, and directions more on the edge followed by outliers. Mental-level dynamics are arguably quite similar, and it can be suggested are best modeled by such a distribution (Salkind, 2007). Mental-level (M) modulation (mod) can hence be described by the following equation-segment:

$$M: Sig * mod\ (G)$$

The physical, the vital, and the mental levels are orientations in which patterns of perceiving, being, behaving are set in their ways. Each pattern has its purpose and its limitation and it can be argued that being able to learn from each orientation and yet being able to move beyond that, is the next logical step in any developmental model. The integral level hence, is about being able to leverage each of the patterns that naturally arise at the three preceding levels at will, and about further, being able to integrate these and arrive at new ways of perceiving and being.

Mathematically such behavior may be represented as being an integrative function ($\int x$) where 'x' is the ability to move between the patterns emanating from G, e, π, at will, represented by $\overline{G, e, \pi}$. Integral-level (I) modulation (mod) of uniqueness (Sig) can hence be represented by the following equation-segment:

$$I\colon\ Sig * mod\ (\int \overline{G, e, \pi}\ )$$

The condition of overcoming any fixed and limiting patterns is the prerequisite for the emergence of 'Force' or for entering into the force-level. At this level the uniqueness behind the particular development being considered can emerge in its purity and become a truly creative dynamic. This aspect of creativity that is in a sense not bound by circumstance may be represented by the constant 'c', the speed of light in a vacuum (Perkowitz, 2011) which is an upper limit of the layer that systems practically operate in. Force-level (F) modulation (mod) of uniqueness (Sig) can hence be represented by the following equation-segment:

$$F\colon\ Sig\ * mod\ (c)$$

Once the signature of an organization arises and continues to exercise itself in its purity, it achieves contextual-mastery (C) and is able to exercise itself as though the context it is acting in, that can vary in scale and complexity, were all of the same substance as itself. This equality may be represented by the integrative function '$\int = 1$'. The equation-segment that notates this contextual-level (C) modulation (mod) applied to organizational uniqueness (Sig) is hence:

$$C\colon\ Sig * mod\ (\int\ = 1)$$

Piecing all the equation-segments together the equation for the emergence of uniqueness (Sig_E), where 'X' can be any of the discussed modulations at the respective development-model levels (P, V, M, I, F, C), is hence summarized by Equation 7.1:

$$Sig_E = X \left\lfloor \begin{array}{c} C: Sig * mod\ (\int = 1) \\ F: Sig\ mod\ (c) \\ I: Sig\ mod\ (\int \overline{G, e, \pi}\ ) \\ M: Sig * mod\ (G) \\ V: Sig * mod\ (e) \\ P: Sig * mod\ (\pi) \end{array} \right\rfloor$$

Eq 7.1: Emergence of Uniqueness

The power of virtual worlds in engaging people and promoting learning as described in "Why Virtual World Matter" (Thomas & Brown, 2009) can be seen as parallel to the emergence of uniqueness as described by Equation 7.1. As described by the authors there is a process of selectively leaving behind part of oneself (movement through the P, V, M levels) to recreate one's identity (movement through level I), and then engaging with others in a shared discourse and culture (movement through levels F, C), which becomes very meaningful for people.

Chapter 8: Varying Culture of Organizations

Having defined an equation for the uniqueness of an organization it now becomes possible to also derive equations for the culture of an organization. Understanding culture in terms of its ability to be a foundation of innovation is important since some cultures cause stagnation while some cause innovation. Useful approaches to and frameworks on innovation at varying levels of organizational complexity are covered in some works of Sri Aurobindo (Sri Aurobindo, 1971). Additionally, works such as Ogbonna & Harris (Ogbonna & Harris, 2000) on the links between leadership, performance, and culture suggest that it is only particular kinds of cultures, those that promote innovativeness as opposed to bureaucracy, for example, where performance improves.

The essential variables in the equation for uniqueness are the X and Y elements (as described in Chapter 6), where X is the primary element that defines that organization's uniqueness and Y are the secondary elements that support or nuance the primary element. X is an element of the X_{Set}, which is the set comprised of elements from $S_{System_{Pr}}, S_{System_P}, S_{System_K}, and\ S_{System_N}$. The Y elements are derived from the Y_{Set} which is also a set comprised of elements from $S_{System_{Pr}}, S_{System_P}, S_{System_K}, and\ S_{System_N}$.

In thinking about variation in culture derived from the same primary element, the variable factor is hence the number of elements in the Y_{Set}. The hypothesis is that the more the number of secondary elements, the more diverse and hence the more stable and innovative the organization will be. If an organization is thought of as having multiple levels, then an equation, Equation 8.1, which will define a monoculture is hence the following:

$$Monoculture: elements_{Y_{set_{(n)}}} = elements_{Y_{set_{(1)}}}$$

Eq 8.1: Monoculture

What this equation is basically stating is that at some level 'n' the number and make-up of elements are exactly the same as the initial seed-state from which the organization culture emerged. Such an organization may do well in a very specific circumstance, and probably the one it was created for. However the reality is that change is the only constant and therefore when faced with a different set of circumstances this organization will not have the source of innovation to allow it to respond differently. It will likely destabilize and go extinct.

By contrast, the equation, Equation 8.2, for a diverse culture is:

$$Diverse\ Culture: elements_{Y_{set_{(n)}}} \propto (1+g)^n$$

Eq 8.2: Diverse Culture

In this equation the number and make-up of elements at some level 'n' in the organization is different than any level that precedes it. The difference is related to an exponential function where 'g' is a growth factor that will increase, likely linearly, as the organization gets more and more complex, and 'n' is the level down in the organization starting from level '1' which is the executive level. The diverse culture will have many more sources of innovation when faced with change because of the plethora of secondary elements that support the primary element.

Chapter 9: Inherent Dynamics of Any System

So far the inherent innovation that exists at the system level and summarized by the nature of a point has been considered. Further, how this deep fount of innovation is present everywhere and how sets that make more practical the range of creative forces available in each of the four components of a point; have also been considered. These architectural forces further define the possibility inherent in any system. Leveraging these sets of forces an equation for the uniqueness of an organization, regardless of scale, was arrived at. Further, it was suggested what made for an innovative versus a stagnant culture.

In some sense the precipitation of innovation from the barely perceptible nature of the ubiquitous point, to how this reveals a play of forces, to how organizations take their seed and grow from these forces, has been traced.

It is useful to now turn full-circle to return to the initial orientations that allowed so much to be suggested about the nature of innovation in the first place. It is useful to look deeper into the nature of the physical, the vital, the mental, and the integral, and to derive equations that in effect will provide further insight into the dynamics of innovation inherent in these orientations.

In Chapter 4 the negative and positive sides of the physical, the vital, and the mental were suggested in the first set of equations that approximated the conditions of Stagnation, Stability, Entropy, Energy, Fragmentation, and Sustainability.

Here a deeper look at the dynamics inherent in each of these orientations that can allow the shift from the negative to positive, or in other words, that reveal the process of innovation in these fundamental orientations is taken.

Hence, starting with the physical, an equation, Equation 9.1, is summarized as:

$$Physical = \begin{bmatrix} M_3 \rightarrow System_{P_T} \\ (\uparrow F \rightarrow I) \\ M_2 \rightarrow S_{Systemi_{P_T}} \\ (\uparrow Sig \rightarrow F) \\ M_1 \rightarrow Sig_P \\ (\uparrow > P_{P)} \\ U \rightarrow Physical_U \end{bmatrix} TC \rightarrow Physical_T$$

$$Where \begin{bmatrix} Physical_U \ni [inertia, lethargy, status\ quo, \dots] \\ Physical_T \ni [adaptability, durability, strength, \dots] \end{bmatrix}$$

Eq 9.1: Inherent Dynamism in Physical

Essentially this equation is laying out the conditions of moving from the untransformed or negative physical state represented by $Physical_U$ to the transformed or positive physical state represented by $Physical_T$.

The first matrix should be read from the bottom to the top:

$$\begin{bmatrix} M_3 \rightarrow System_{P_T} \\ (\uparrow F \rightarrow I) \\ M_2 \rightarrow S_{System_{P_T}} \\ (\uparrow Sig \rightarrow F) \\ M_1 \rightarrow Sig_P \\ (\uparrow > P_{P)} \\ U \rightarrow Physical_U \end{bmatrix}$$

Hence, at the bottom is the starting point '$U \rightarrow Physical_U$' which identifies the default or untransformed (U) level of the physical. The next row up, $(\uparrow > P_P)$, states that when the patterns of the untransformed physical (P_P) have been overcome (>), movement to the next level (↑) is facilitated. Breaking through to the next level, $M_1 \rightarrow Sig_P$, allows its dynamics to become active. Hence, the signature or uniqueness of the physical (Sig_P) becomes active at meta-level 1 (M_1). As this signature becomes more like a Force ($Sig \rightarrow F$), the conditions for breakthrough (↑) to the next level are achieved. This next level is referred to as meta-level 2 (M_2), and indicates that the architectural forces represented by the set of system-presence ($S_{system_{P_r}}$) have become more consciously active. When this Force becomes Integral ($F \rightarrow I$) then the conditions for breakthrough (↑) to the next level are achieved. The next level is notated as M_3 for meta-level 3, and the dynamics here indicate that the equation for system-presence becomes active. Becoming active basically means that the respective meta-level dynamic begins to act at the once 'untransformed' level (U) further modifying it. Modification or transformation began when M_1 became active. Transformation is accelerated when M_2 becomes active, and even further accelerated when M_3 becomes active. Note that the matrix is essentially semiotic in that adjacent levels exist in paired relationships and inter-related functionality (Usó-Doménech et al, 2016).

The notion of meta-layers is being explored by contemporary physicists and Erwin Laszlo in his book Self-Actualizing Cosmos (Laszlo, 2014) summarizes some of these developments: "Physicists describe the domain that underlies and embeds the particles, fields, and forces of the universe variously as quantum vacuum, physical spacetime, nuether, zero-point field, grand-unified field, cosmic plenum, or string-net liquid." Note that 'nuether' refers to a sub-quantum level of reality (Pearson, 1997). Laszlo goes on to describe a revolutionary discovery by Nika Arkani-Hamed of Princeton's Institute of Advanced Study, of a geometrical object, the amplituhedron (Arkani-Hamid et al, 2012), which is not in space-time but governs space-time so that it "appears that spatio-temporal phenomena are the consequence of geometrical relationships in a deeper dimension of physical reality". A deeper dimension of a physical layer suggests synonymity with a meta-layer.

The rate of the transformation can be better envisioned when considering action of the Transformation Circle, or TC. The TC can be thought of as 4 concentric circles, with M_3 at the center. M_3 is surrounded by M_2, which is surrounded by M_1. The outer circle is U. If TC is considered to be a clock, than at time 't = 0', the physical' can be thought of as being entirely in U. The clock starts ticking only when some initial patterns P_P are overcome $(> P_P)$. From this point on as time proceeds the conditions for breakthrough become riper, and a sinusoidal wave begins to integrate more of the concentric circles together. The sinusoid wave (sin) is itself modulated by an euler function, e^x, where 'x' is determined by the strength to overcome patterns (↑) which will likely vary over time but will likely tend to be positive once the clock has started ticking because of the joy experienced with progressive movement. Being that the limit is the outer boundary of the concentric circles, there is further modulation by π until the 4 concentric circles have been integrated. TC, hence, may be represented by Equation 9.2:

$$TC \equiv (> P_P) \rightarrow \text{mod}\,(\sin, e^x, \pi)$$

Eq 9.2: Transformation Circle

Hence, the initial nature of the physical that may be characterized by the set comprising of elements such as inertia, lethargy, acceptance of the status quo, amongst other such elements

($Physical_U \ni [inertia, lethargy, status\ quo, \ldots]$), transforms into a physical more characterized by elements such as adaptability, durability, strength, and so on ($Physical_T \ni [adaptability, durability, strength, \ldots$). This transformation represents the inherent innovation-dynamic within the Physical.

Similarly, the equation for the 'Vital', Equation 9.3, also shows the built-in transformation that represents the innovation-dynamic within the vital:

$$Vital = \begin{bmatrix} M_3 \rightarrow System_P \\ (\uparrow F \rightarrow I) \\ M_2 \rightarrow S_{Systemp} \\ (\uparrow Sig \rightarrow F) \\ M_1 \rightarrow Sig_V \\ (\uparrow > P_{V)} \\ U \rightarrow Vital_U \end{bmatrix} TC \rightarrow Vital_T\ ,$$

$$Where \begin{bmatrix} Vital_U \ni [aggression, self\ centeredness, exploitati\square n, \ldots] \\ Vital_T \ni [energy, support, adventure, enthusiasm, \ldots] \end{bmatrix}$$

Eq 9.3: Inherent Dynamism of Vital

The equation for the 'Mental', Equation 9.4, is similarly summarized as:

$$Mental = \begin{bmatrix} M_3 \rightarrow System_S \\ (\uparrow F \rightarrow I) \\ M_2 \rightarrow S_{Systems_S} \\ (\uparrow Sig \rightarrow F) \\ M_1 \rightarrow Sig_M \\ (\uparrow > P_{M)} \\ U \rightarrow Mental_U \end{bmatrix} TC \rightarrow Mental_T$$

$$Where \begin{bmatrix} Mental_U \ni [fixation, fundamentalism, fragmentation, \ldots] \\ Mental_T \ni [understanding, imagination, inspiration, \ldots] \end{bmatrix}$$

Eq 9.4: Inherent Dynamism of Mental

The equation for the 'Integral', Equation 9.5, is similarly summarized as:

$$Integral = \begin{bmatrix} M_3 \rightarrow System_N \\ (\uparrow F \rightarrow I) \\ M_2 \rightarrow S_{System_N} \\ (\uparrow Sig \rightarrow F) \\ M_1 \rightarrow Sig_I \\ (\uparrow > P_{I)} \\ U \rightarrow Integral_U \end{bmatrix} TC \rightarrow Integral_T$$

$$Where \begin{bmatrix} Integral_U \ni [possession, usurpation, hidden\ agendas, \ldots] \\ Integral_T \ni [appreciation, shift\ POV, MPV, synthesis, \ldots] \end{bmatrix}$$

Eq 9.5: Inherent Dynamism of Integral

The preceding equations can be generalized by Equation 9.6:

$$Innovation_{orientation-x} = \begin{bmatrix} M_3 \rightarrow System_X \\ (\uparrow F \rightarrow I) \\ M_2 \rightarrow S_{System_X} \\ (\uparrow Sig \rightarrow F) \\ M_1 \rightarrow Sig_x \\ (\uparrow > P_{x)} \\ U \rightarrow x_U \end{bmatrix} TC \rightarrow x_T\ , where \begin{bmatrix} x_U \ni [\ldots] \\ x_T \ni [\ldots] \end{bmatrix}$$

Eq 9.6: Generalized Equation of Innovation

In this generalized equation, $Innovation_{orientation-x}$, refers to the inherent innovation within a specific orientation. Orientation refers to the physical, the vital, the mental, or the integral.

Further, the notion of a core-matrix that will be referred to in subsequent chapter can be summarized by the following equation, Equation 9.7:

$$Core\_matrix = \begin{bmatrix} M_3 \rightarrow System_X \\ (\uparrow F \rightarrow I) \\ M_2 \rightarrow S_{System_X} \\ (\uparrow Sig \rightarrow F) \\ M_1 \rightarrow Sig_x \\ (\uparrow > P_{x)} \\ U \rightarrow x_U \end{bmatrix}$$

Eq 9.7: Core Matrix

One of the corollaries of the Generalized Equation of Innovation is that if the source of innovation is more influenced by a meta-level there will be simultaneity of innovation that becomes apparent at U. The higher the meta-level, the more likely that this simultaneity will be wider spread. In his book Where Good Ideas Come From: The Natural History of Innovation (Johnson, 2010) Johnson states: "A brilliant idea occurs to a scientist or inventor somewhere in the world, and he goes public with his remarkable finding, only to discover that three other minds had independently come up with the same idea in the past year." He refers to an essay "Are Inventions Inevitable" (Ogburn & Thomas, 1922) which uncovered 148 instances of similar yet independent innovation, most of them occurring within the same decade. Some examples include sunspots that were simultaneously discovered in 1611 by four scientists in four different countries, and the law of conservation of energy that was formulated separately four times in the late 1840s, amongst numerous other example.

Chapter 10: Equations for Stagnation and Dynamic Growth

Considering the equations for the Physical, the Vital, the Mental, and the Integral, presented in the previous chapter, the implicit reality of innovation present within them may become apparent. This sense of innovation is embedded within these fundamental orientations. But as the Transformation Circle, TC, indicates, each of the respective untransformed sets transforms to each of the respective transformed sets only if TC becomes active.

An easy way, hence, to characterize stagnation (or the lack of innovation) and dynamic growth (the presence of innovation) is through the following generalized equations where 'x' can be thought as the Physical, the Vital, the Mental, or the Integral.

The generalized equation for Stagnation, Equation 10.1, hence is:

$$\frac{d\,(TC_x)}{dt} \leq 0$$

Eq 10.1: Generalized Equation for Stagnation

The generalized equation for Dynamic Growth, Equation 10.2, is:

$$\frac{d\,(TC_x)}{dt} > 0$$

Eq 10.2: Generalized Equation for Dynamic Growth

As a corollary it may also be suggested that rather than being universal, the second law of thermodynamics is perhaps active under conditions of stagnation, as in

Equation 10.3. Of course there is currently no proof of this.

Hence,

IF $\frac{d\,(TC_x)}{dt} \leq 0$ THEN dS >= 0 (where S is Entropy)

Eq 10.3: Suggested Equation for Entropy

If, as being suggested in the mathematical model being derived here that there is a fundamental process for the precipitation of innovation, that there are fundamental orientations implicit with dynamism, and further that there are conditions for growth and stagnation, then the notion of randomness in CAS will have to be revisited. This is done in the next Chapter.

Chapter 11: Qualified Determinism of Complex Adaptive Systems

The model and mathematics enumerated in this work hence, suggests though that the randomness being projected by CAS scientists (Dodder & Dare, 2000) does not really exist. Let us look at aspects of the mathematics to substantiate this.

A new function, Dynamic Interaction (DI) that has a 'vertical' and a 'horizontal' component is introduced here. The vertical component is designated as DI_V and the horizontal component as DI_H. Several equations to capture the inherent dynamism that exists in each orientation or state have already been derived in Chapter 9. These included equations for the dynamism in the physical, the vital, the mental, and the integral. The derived equations propose a model to give insight into how innovation occurs by changing the fundamental states that an organization is subject to. CAS scientists such as Prigogine (Prigogine, 1977) and others are proposing that a system can bifurcate in unpredictable ways to create an emergent property that cannot be predicted. DI is going to propose that in fact there is a 'qualified determinism' as opposed to randomness that occurs.

This qualified determinism is the result of the relative strengths of the levels within the core-matrix identified in the derived equations and summarized by Equation 9.7. Hence, the generalized core-matrix, already is the following:

$$\begin{bmatrix} M_3 \rightarrow System_X \\ (\uparrow F \rightarrow I) \\ M_2 \rightarrow S_{System_X} \\ (\uparrow Sig \rightarrow F) \\ M_1 \rightarrow Sig_x \\ (\uparrow > P_{x)} \\ U \rightarrow x_U \end{bmatrix}$$

The application of the vertical component of the new function being proposed, DI_V, to this core matrix will yield the nature or 'strength' of the state (x) or orientation under consideration. If the untransformed or U layer is strongest, implying that the habitual patterns that keep an organization locked into its untransformed way of operation are still very active, then the nature of the output of DI_V, notated by x-state, will be x_U. If the habitual patterns have been overcome then the strength of the x-state increases since it is the dynamics of M_1 or Sig_x that are now active. In this case the x-state will be Sig_x. If the unique 'signature' has become a 'force', then the conditions for activation of M_2 have been put in place and the x-state will be even higher, S_{System_X}. The architectural forces active in M_2 are by definition more powerful than Sig_x that is a derivation of a set of such architectural forces. If the 'force' so acting becomes impersonal so that an organizational ego-state is overcome, then the x-state will have the most strength and is characterized by $System_X$ active at M_3. Hence, DI_V applied to a core-matrix will yield the 'strength' in terms of the x-dynamic that is active. This is illustrated by the following equation, Equation 11.1 which can be considered to be a deductive proof in the context of this model:

$$DI_V \begin{bmatrix} M_3 \rightarrow System_X \\ (\uparrow F \rightarrow I) \\ M_2 \rightarrow S_{System_X} \\ (\uparrow Sig \rightarrow F) \\ M_1 \rightarrow Sig_x \\ (\uparrow > P_{x)} \\ U \rightarrow x_U \end{bmatrix} =>$$

$$\text{x-state} \in \left(x_U, Sig_x, S_{System_X}, System_X\right)$$

$$Where: Strength(System_X) > Strength(S_{System_X}) > Strength(Sig_x) > Strength(x_U)$$

Eq 11.1: Illustrating Action of Dynamic Interaction – vertical component

What is to be noted here is that while the action of DI_V yields a relative strength and therefore a 'single' value for the core- or x-matrix under consideration yet each x-matrix in itself could have an infinite number of possibilities. This should be clear in looking at how x_U, Sig_x, S_{System_X}, and $System_X$, were initially defined.

Hence, taking the example where x = physical:

$$Physical_U \ni [inertia, lethargy, status\ quo, \ldots]$$

As can be seen $Physical_U$, defined in Chapter 9, is already an infinite set with qualities similar to the ones already specified.

Similarly, Sig_P, defined in Chapter 6, also has an infinite variation:

$$Sig_P = Xa + \overline{Yb_{0-n}} \quad where \begin{bmatrix} X \in [S_{System_{Pr}}] \\ Y \in [S_{System_{Pr}}, S_{System_P}, S_{System_K}, S_{System_N}] \\ a, b \ are \ integers; a > b \end{bmatrix}$$

$S_{System_{Pr}}$, defined in Chapter 5, is also an infinite set with forces of the nature specified in the following equation:

$$S_{System_{Pr}} \ni [Service, Perfection, Diligence, Perseverance, \ldots]$$

And recall that in Chapter 3, $System_{Pr}$ has been defined as:

$$System_{Pr} \equiv TS_{0 \to N} \begin{bmatrix} \begin{matrix} Presence \\ \downarrow \\ Opportunity \end{matrix} \begin{bmatrix} P_L & \to & V_L \\ V_L & \to & M_L \end{bmatrix} \ \& \ Container_{\begin{bmatrix} System_P \\ System_K \\ System_N \end{bmatrix}} \end{bmatrix}$$

So in essence DI_V is really giving us a summary assessment of the 'level' of the x-matrix under consideration with all its infinite potentiality. An example will follow shortly.

The other component of DI, as suggested earlier in this chapter, is the horizontal component, DI_H. Just as DI_V yields a summary assessment of the level that an x-matrix is operating at, similarly DI_H yields a summary assessment of the direction that a system or organization under consideration is going to continue its development in considering the physical, the vital, the mental, and the integral orientations to be the choices.

Assuming that any organization or system is inherently unique, as this mathematical model proposes, and assuming that the infinite sets of x_U and S_{System_X} applied across the physical, vital, mental, and integral orientations respectively will account for any state that an organization can experience, then at a certain point in time any organization under consideration is going to have a direction-bias in one of the possible physical, vital, mental, or integral directions. Hence, DI_H will yield the summary direction that is going to lead an organization into its future given the current states active in it.

This summary direction is going to be yielded by considering the relative strengths of the separate core x-matrices – the physical, the vital, the mental, the integral - under consideration. The assumption is that there will be one core-matrix that will be stronger than the others.

Hence, as an example, first applying DI_V across all four x-matrices may, for example, yield the following results, with the strongest level within each x-matrix highlighted and bolded:

$$\begin{bmatrix} System_{Pr} \\ \boldsymbol{S_{System_{Pr}}} \\ Sig_P \\ Physical_U \end{bmatrix} \begin{bmatrix} System_P \\ S_{System_P} \\ Sig_V \\ \boldsymbol{Vital_U} \end{bmatrix} \begin{bmatrix} System_K \\ S_{System_K} \\ Sig_M \\ \boldsymbol{Mental_U} \end{bmatrix} \begin{bmatrix} System_N \\ S_{System_N} \\ \boldsymbol{Sig_I} \\ Integral_U \end{bmatrix}$$

Since by definition the strength of $System_x$ is greater than S_{System_x}, which is greater than Sig_x, which is greater than x_U, applying DI_H, as in Equation 11.2, across these x-matrices, as in the example following it will then yield the strongest direction, which in this example is the Physical:

$$DI_H\left(\begin{bmatrix} System_{Pr} \\ S_{System_{Pr}} \\ Sig_P \\ Physical_U \end{bmatrix} \begin{bmatrix} System_P \\ S_{System_P} \\ Sig_V \\ Vital_U \end{bmatrix} \begin{bmatrix} System_K \\ S_{System_K} \\ Sig_M \\ Mental_U \end{bmatrix} \begin{bmatrix} System_N \\ S_{System_N} \\ Sig_I \\ Integral_U \end{bmatrix}\right) = Orientation_{Strongest}$$

Eq 11.2: Illustrating Action of Dynamic Interaction - horizontal component

Example:

$$DI_H\left(\begin{bmatrix} System_{Pr} \\ \mathbf{S_{System_{Pr}}} \\ Sig_P \\ Physical_U \end{bmatrix} \begin{bmatrix} System_P \\ S_{System_P} \\ Sig_V \\ \mathbf{Vital_U} \end{bmatrix} \begin{bmatrix} System_K \\ S_{System_K} \\ Sig_M \\ \mathbf{Mental_U} \end{bmatrix} \begin{bmatrix} System_N \\ S_{System_N} \\ \mathbf{Sig_I} \\ Integral_U \end{bmatrix}\right) = Physical$$

Hence, DI function will yield the following organizational direction, as in Equation 11.3, where 'x_matrix' is used interchangeably with 'orientation':

$$Org\_Dir = DI\left(\begin{matrix} \begin{bmatrix} M_3 \rightarrow System_{Pr} \\ (\uparrow F \rightarrow I) \\ M_2 \rightarrow S_{System_{Pr}} \\ (\uparrow Sig \rightarrow F) \\ M_1 \rightarrow Sig_P \\ (\uparrow > P_{P)} \\ U \rightarrow Physical_U \end{bmatrix} \begin{bmatrix} M_3 \rightarrow System_P \\ (\uparrow F \rightarrow I) \\ M_2 \rightarrow S_{System_P} \\ (\uparrow Sig \rightarrow F) \\ M_1 \rightarrow Sig_V \\ (\uparrow > P_{V)} \\ U \rightarrow Vital_U \end{bmatrix} \\ \begin{bmatrix} M_3 \rightarrow System_S \\ (\uparrow F \rightarrow I) \\ M_2 \rightarrow S_{System_S} \\ (\uparrow Sig \rightarrow F) \\ M_1 \rightarrow Sig_M \\ (\uparrow > P_{M)} \\ U \rightarrow Mental_U \end{bmatrix} \begin{bmatrix} M_3 \rightarrow System_N \\ (\uparrow F \rightarrow I) \\ M_2 \rightarrow S_{System_N} \\ (\uparrow Sig \rightarrow F) \\ M_1 \rightarrow Sig_I \\ (\uparrow > P_{I)} \\ U \rightarrow Integral_U \end{bmatrix} \end{matrix}\right) \rightarrow x\_matrix_{strongest} @ level_{strongest}$$

Eq 11.3: Organizational Direction

Generalizing, as in Equation 4.9.4, where Org_Dir is organizational direction:

$$Org\_Dir = DI\left(\begin{bmatrix} M_3 \rightarrow System_X \\ (\uparrow F \rightarrow I) \\ M_2 \rightarrow S_{System_X} \\ (\uparrow Sig \rightarrow F) \\ M_1 \rightarrow Sig_x \\ (\uparrow > P_{P)} \\ U \rightarrow x_U \end{bmatrix}_{x=p,v,m,i}\right) \rightarrow$$

$$x\_matrix_{strongest} \text{ @ } level_{strongest}$$

Eq 11.4: Generalized Equation for Organizational Direction

Hence, this mathematical model is suggesting that any situation, rather than having a random outcome, has a 'qualified deterministic' outcome. In the introduction to his book "Where is Science Going?" (Planck, 1933), James Murphy points out that the reason Planck spent so much of his time giving lectures on causation was because of the trend of physicists at the time, which has continued to the modern day, to overthrowing the principle of causation following the development of quantum theory, which he felt was misplaced. "Planck would claim", he wrote, "and so would Einstein, that it is not the principle of causation itself which has broken down in modern physics, but rather the traditional formulation of it." Murphy also quotes James Jeans (Jeans, 1932) to suggest the issue associated with causation and determinism: "Einstein showed in 1917 that the theory founded by Planck appeared, at first sight at least, to entail consequences far more revolutionary than mere discontinuity", and here he is referring to the finding that radiant energy is not emitted in a continuous flow, but in integral quantities, or quanta, which can be expressed in integral numbers. Continuing: "It appeared to dethrone the law of causation from the position it had therefore held as guiding the course of the natural world. The old science had confidently proclaimed that nature could follow only one road, the road which was mapped out from the beginning of time to its end by the continuous chain of cause and effect; state A was inevitably succeeded by state B. So far the new science has only been able to say that state A may be followed by state B or C or D or by innumerable other states. It can, it is true, say that B is more likely than C, C than D, and so on; it can even specify the relative probabilities of B, C, and D. But, just because it has to speak in terms of probabilities, it cannot speak with certainty which state will follow which; this is a matter which lies on the knees of the gods – whatever gods there may be."

While under the apparent dynamics at the quantum level there may appear to be randomness and a dethroning of the principle of causation, the notion of a multiplicity of levels, each having its impact on the strength of an orientation and further on the consequent direction from a multiplicity of possible orientations, is being suggested here as determining the direction of any CAS, while still allowing infinite variation in the details that may define its. Hence, the positions of Planck and Einstein are vindicated when considering Equation 11.4.

Further, assuming any CAS where multiple elements are active, connected, interdependent, and emergent, it may be possible to understand, through application of calculus, as to which level is the source for change.

Hence, where N may be source of change, the rate of change of N will resolve into one of P_U, V_U, M_U, I_U, P_T, V_T, M_T, or I_T. This may be summarized by Equation 11.5, where y is either U or T:

$$\frac{dN}{dt} \rightarrow \begin{bmatrix} P_U & P_T \\ V_U & V_T \\ M_U & M_T \\ I_U & I_T \end{bmatrix} \rightarrow x_y, where\ y \in (U,T)$$

Eq 11.5: Establishing the Nature of the Change

If T, implying that the action of one of the meta-levels has caused transformation, then application of one of the following integrals will determine which level is the likely source for change.

Hence, for M_1 if the integral of $\frac{\partial(x_U \rightarrow x_T)}{\partial t}$ across a limited area 'a' in the vicinity of the change, is greater than some threshold value $Threshold_{Signature}$, then the signature dynamics are likely the source of change. This is summarized by Equation 11.6:

$$\int_0^a \frac{\partial(x_U \rightarrow x_T)}{\partial t} > Threshold_{Signature}$$

Eq 11.6: Signature Dynamics as the Source of Change

For M_2 if the integral of $\frac{\partial(x_U \rightarrow x_T)}{\partial t}$ across a larger area 'b' extending beyond the vicinity of the change, is greater than some threshold value $Threshold_{Architectural\ Forces}$, then the architectural forces are likely the source of change. This is summarized by Equation 11.7:

$$\int_0^b \frac{\partial(x_U \rightarrow x_T)}{\partial t} > Threshold_{ArchitecturalForces}$$

Eq 11.7: Architectural Forces as the Source of Change

For M_3 if the double integral of $\frac{\partial(x_U \rightarrow x_T)}{\partial t}$ across the CAS specified by 'A', and across some time 't', is greater than some threshold value $Threshold_{System\ Property}$, then the system properties are likely the source of change. This is summarized by Equation 11.8:

$$\int_0^t \int_0^A \frac{\partial(x_U \rightarrow x_T)}{\partial t} > Threshold_{SystemProperty}$$

Eq 11.8: System Properties as the Source of Change

Chapter 12: Framing Organizational Transitions at Layer U

Dissipative Structures exist as stabilities far from equilibrium. They are characterized by formation of complex structures that tend to exhibit long-range correlations between interacting particles. Mathematically, dissipative structures can be thought of as consisting of states x(t), inputs u(t), and outputs y(t) such that the energy stored in the emerging structure V(x), can never exceed the energy supplied to the system, uy (Prigogine, 1967).

This can be represented by the inequality (Prigogine-Dissipative Inequality):

$$\frac{dV(x(t))}{dt} \leq u(t).y(t)$$

It has been suggested that V(x) may play the role of Lyapunov functions that function to stabilize systems (Willems, 1998).

These dissipative structures can be thought of as the frame within which the many unique organizations that surface or manifest exist.

Further, it may be suggested that there are different kinds of dissipative structures depending on what kind of organizational transition is seeking to express itself. The 'context' of the dissipative structure being created hence becomes important. Since there are many steps in the journey from x_U to x_T this in effect may frame the nature of the transition that is taking place. The general equation for the emergence of uniqueness, Equation 7.1, derived in Chapter 7, provides a framework for the possible steps in such a transition. Recalling:

$$Sig_E = X \begin{vmatrix} C: Sig * mod\ (\int = 1) \\ F: Sig\ mod\ (c) \\ I: Sig\ mod\ (\int \overline{G, e, \pi}\ ) \\ M: Sig * mod\ (G) \\ V: Sig * mod\ (e) \\ P: Sig * mod\ (\pi) \end{vmatrix}$$

Sig_E is suggested as the model for emergence of the physical, the vital, the mental, and the integral orientations. At the very least therefore there would be 24 different storage functions, V(x), that would be possible and each would have a different nature. These 24 possibilities are the product of the six states in Sig_E times the four orientations. Orientation may be represented by O, and the states within these by S. Hence:

$$OS \in \begin{bmatrix} System_{Pr} \\ S_{System_{Pr}} \\ Sig_P \\ Physical_{Um} \\ Physical_{Uv} \\ Physical_{Up} \end{bmatrix} \begin{bmatrix} System_P \\ S_{System_P} \\ Sig_V \\ Vital_{Um} \\ Vital_{Uv} \\ Vital_{Up} \end{bmatrix} \begin{bmatrix} System_K \\ S_{System_K} \\ Sig_M \\ Mental_{Um} \\ Mental_{Uv} \\ Mental_{Up} \end{bmatrix} \begin{bmatrix} System_N \\ S_{System_N} \\ Sig_I \\ Integral_{Um} \\ Integral_{Uv} \\ Integral_{Up} \end{bmatrix}$$

In this case the basic dissipative inequality may be enhanced, as suggested in Equation 12.1, to:

$$\frac{dV_{OS}(x(t))}{dt} \leq u(t).y(t)$$

Eq 12.1: Dissipative Inequality Generalized for Orientation-State

Further, recall the innovative dynamism in our system captured by the general equation, Equation 9.6, derived in Chapter 9:

$$Innovation_{orientation-x} = \begin{bmatrix} M_3 \rightarrow System_X \\ (\uparrow F \rightarrow I) \\ M_2 \rightarrow S_{System_X} \\ (\uparrow Sig \rightarrow F) \\ M_1 \rightarrow Sig_x \\ (\uparrow > P_{x)} \\ U \rightarrow x_U \end{bmatrix} TC \rightarrow x_T \text{ , } where \begin{bmatrix} x_U \ni [\ldots] \\ x_T \ni [\ldots] \end{bmatrix}$$

The essential transition in this innovative dynamism is occurring from x_U to x_T. As an organization approaches x_T there will be a higher degree of information and order associated with it and it may be suggested that in its transition from x_U to x_T the storage energy $V_{OS}(x)$ will simultaneously increase. This may be expressed by Equation 12.2 as:

$$\frac{dV_{OS}(x_U(t))}{dt} < \frac{dV_{OS}(x_T(t))}{dt}$$

Eq 12.2: Transition Inequality from x_U to x_T

Note that in replacing x(t) in the original form of the inequality by $x_g(t)$ where g is u or t, allows for multi-layer causal dynamics that is absent from a framing where only x(t) is considered. Philosophically this framing transcends the limit to explain phenomena at surface-layer U by mechanics at the same layer only.

Further, and keeping in mind the six states of progression identified by Sig_E, it may be suggested that in its journey from x_U to x_T progressively more complex storage functions mapped to each of these states are reached and traversed, as identified by the multiple inequalities depicted in Equation 12.3:

$$\frac{dV_{OS}(x_U(t))}{dt} \rightarrow \frac{dV_{OS}(x_T(t))}{dt}:$$

$$\frac{dV_{OS}(x_P(t))}{dt} < \frac{dV_{OS}(x_V(t))}{dt} < \frac{dV_{OS}(x_M(t))}{dt} < \frac{dV_{OS}(x_I(t))}{dt} < \frac{dV_{OS}(x_F(t))}{dt} < \frac{dV_{OS}(x_C(t))}{dt}$$

Eq 12.3: Transition Inequality Depicting Multi-Layer Causal Dynamics

Chapter 13: Framing and Modeling Shifts in Innovation at Layer U

If innovation is assumed as generally being the breaking of certain existing patterns and the formation of new ones then it is should be possible to leverage Turing's activator-inhibitor equations (Turing, 1952) to frame shifts in innovation at the surface level, U. The notion of innovation as being the adoption of new patterns is consistent with Cognitive Psychologist Sternberg's model of creativity (Sternberg, 2012). In the 'Assessment of Creativity: An Investment Based Approach' he writes: "Creativity is a habit. Behind all innovations one finds creativity, so innovations arise from a habit. When I speak of a habit, I refer to 'an acquired behavior pattern regularly followed until it has become almost involuntary'. That is, creativity becomes a way of life that one regularly utilizes so that one is hardly aware one is engaging in it."

Turing suggested the use of an activator (u) and an inhibitor (v). The activator stimulates production of both entities, whereas the inhibitor inhibits it. Applying his equations u may be thought of as being the 'new innovative pattern' and v as being the 'existing pattern(s)'. The relevant system of equations, where the diffusion rates for u and v are D_u and D_v is then:

$$\frac{\partial u}{\partial t} = f(u,v) + D_u \nabla^2 u$$
$$\frac{\partial v}{\partial t} = g(u,v) + D_v \nabla^2 v$$

In light of the preceding discussion it may be suggested that a shift between any of the ordered states at layer U may be 'framed' rather than 'modeled' by the preceding Turing equations. 'Modeling' will require integration or a mapping of the action of the meta-layers into the shifts being observed and is intended to capture the fuller set of causal dynamics. 'Framing' is a relatively static orientation that only captures the observed surface effects.

Recall the equation of innovative dynamism, Equation 9.6 derived in Chapter 9:

$$Innovation_{orientation-x} = \begin{bmatrix} M_3 \rightarrow System_X \\ (\uparrow F \rightarrow I) \\ M_2 \rightarrow S_{System_X} \\ (\uparrow Sig \rightarrow F) \\ M_1 \rightarrow Sig_x \\ (\uparrow > P_{x)} \\ U \rightarrow x_U \end{bmatrix} TC \rightarrow x_T \text{ , } where \begin{bmatrix} x_U \ni [\ldots] \\ x_T \ni [\ldots] \end{bmatrix}$$

It may be suggested that the general transition from x_U to x_T can be framed at the layer U by the Turing activator-inhibitor equations. Hence, replacing the 'u' or activator, or new innovative pattern by x_T, and the 'v' or inhibitor, or existing pattern by x_U an equation, Equation 13.1, to frame shift in innovation at U is arrived at:

$$Shift_{innovation} = \left(\frac{\partial x_T}{\partial t} = f(x_T, x_U) + D_{x_T}\nabla^2 x_T \quad , \quad \frac{\partial x_U}{\partial t} = g(x_T, x_U) + D_{x_U}\nabla^2 x_U\right)$$

Eq 13.1: Shift in Innovation Leveraging Turing Activator-Inhibitor Equations

Note the similar philosophical point here as in the previous chapter. The Turing activator-inhibitor equations had been defined as operating at the surface layer only. Hence, the cause for the shifts occurring at U are suggested as the actions of u and v which are themselves surface-layer causal-agents. By contrast the modified form that appears in $Shift_{innovation}$ contains in it the action implicit at meta-layers by replacing u and v by x_T and x_U respectively.

Chapter 14: Framing Complexity

There has been much written about the difference between complicated and complex systems (Aral, 2016; Efatmaneshnik et al., 2012). Leveraging off a representative piece, Learning to Live with Complexity that appeared in Harvard Business Review (Sargut & McGrath, 2011), here are some relevant points:

- Simple systems contain few interactions and are predictable
- Complicated systems have many parts but operate in patterned ways
- Complex systems are imbued with features that may operate in patterned ways but whose interactions are continually changing
- Complex systems interact in unexpected ways
- It is harder to make sense of things because the degree of complexity may exceed our cognitive limits
- Analytical tools have not kept up with complexity thinking
- Three properties determine the complexity of an environment. The first, *multiplicity,* refers to the number of potentially interacting elements. The second, *interdependence,* relates to how connected those elements are. The third, *diversity,* has to do with the degree of their heterogeneity. The greater the multiplicity, interdependence, and diversity, the greater the complexity.
- In Complicated systems one can predict the outcome given starting conditions. In complex systems the same initial conditions can create different outcomes depending on interaction of elements in the system and it is difficult to model complex systems because elements interact continuously and unpredictably

The differentiations point to some useful aspects to be considered in a mathematics for CAS. For example, a CAS mathematics will become more robust by considering multiplicity, interdependence, and diversity of a system.

Further, the notion that system complexity may be a function of multiplicity, interdependence, and diversity suggests too that the simple, complicated, and complex systems may be related by multiplicity, interdependence, and diversity. That is, they are all different expressions of CAS, perhaps at different levels of maturity.

Hence, complexity in a system may be thought of as a function of multiplicity (the number of interacting components), interdependence (how connected those elements are), and diversity (the degree of heterogeneity). Summarizing:

$$Complexity_{System} = f\ (Multiplicity, Interdependence, Diversity)$$

Recall that the general scheme of meta-levels is encapsulated by the core-matrix, as stated in Equation 9.7:

$$\begin{bmatrix} M_3 \rightarrow System_X \\ (\uparrow F \rightarrow I) \\ M_2 \rightarrow S_{System_X} \\ (\uparrow Sig \rightarrow F) \\ M_1 \rightarrow Sig_x \\ (\uparrow > P_{x)} \\ U \rightarrow x_U \end{bmatrix}$$

If the untransformed or U layer is strongest, implying that the habitual patterns that keep an organization locked into its untransformed way of operation are still very active, then the nature of the system will be x_U.

With x_U active the complexity of the system can be thought of in the following manner:

- Low diversity: Untransformed physical, vital, mental, and integral elements that are the actors in this system are basically homogeneous since habitual patterns keep behavior locked in place. Taking the example where x = physical, this homogeneity is described in the following manner:

$$Physical_U \ni [inertia, lethargy, status\ quo, \ldots]$$

- Low multiplicity: Being homogeneous the number of interacting components is lower than what it can be.
- Low interdependence: These elements are not sufficiently unique to have become organizing forces, and therefore the level of interdependence will also be low

If the habitual patterns have been overcome the dynamics of M_1 or Sig_x are now active. The dynamics of Sig_X have now become active which opens the possibility to a vast number of heterogeneous elements. These elements are all organized enough that they will cause interdependence. Considering the case where x = physical:

$$Sig_P = Xa + \overline{Yb_{0-n}} \quad where \begin{bmatrix} X \in [S_{System_{Pr}}] \\ Y \in [S_{System_{Pr}}, S_{System_P}, S_{System_K}, S_{System_N}] \\ a, b\ are\ integers; a > b \end{bmatrix}$$

Where $S_{System_{Pr}}$ is an infinite set with forces of the nature specified in the following manner:

$$S_{System_{Pr}} \ni [Service, Perfection, Diligence, Perseverance, \ldots]$$

Hence, the complexity of the system will increase with the increase in multiplicity, interdependence, and diversity.

If the unique 'signature' has become a 'force', then the conditions for activation of M_2 have been put in place and the system state will be S_{System_X}. The architectural forces active in M_2 are by definition more powerful than Sig_x that is a derivation of a set of such architectural forces.

If the 'force' so acting becomes impersonal so that an organizational ego-state is overcome, then the system state will have the most strength and is characterized by $System_X$ active at M_3.

Hence, it is seen that in general as the higher meta-level becomes active, the system becomes more complex as measured by the function $Complexity_{System}$.

But a function DI_V, summarized in Equation 11.1 had already been derived in Chapter 11 that indicates which level in a system is active:

$$
DI_V \begin{bmatrix} M_3 \rightarrow System_X \\ (\uparrow F \rightarrow I) \\ M_2 \rightarrow S_{System_X} \\ (\uparrow Sig \rightarrow F) \\ M_1 \rightarrow Sig_x \\ (\uparrow > P_{x)} \\ U \rightarrow x_U \end{bmatrix} =>
$$

x-state $\in \left(x_U, Sig_x, S_{System_X}, System_X\right)$

$$
Where: Strength(System_X) > Strength(S_{System_X}) > Strength(Sig_x) > Strength(x_U)
$$

As a reminder the application of DI_V will yield the nature or 'strength' of the state (x) or orientation under consideration. If the untransformed or U layer is strongest, implying that the habitual patterns that keep an organization locked into its untransformed way of operation are still very active, then the nature of the output of DI_V, notated by x-state, will be x_U. If the habitual patterns have been overcome then the strength of the x-state increases since it is the dynamics of M_1 or Sig_x that are now active. In this case the x-state will be Sig_x. If the unique 'signature' has become a 'force', then the conditions for activation of M_2 have been put in place and the x-state will be even higher, S_{System_X}. The architectural forces active in M_2 are by definition more powerful than Sig_x that is a derivation of a set of such architectural forces. If the 'force' so acting becomes impersonal so that an organizational ego-state is overcome, then the x-state will have the most strength and is characterized by $System_X$ active at M_3.

Hence, a modified equation, Equation 14.1, for system complexity can be arrived at, where:

$$Complexity_{System} \propto DI_V$$

Eq 14.1: System Complexity

Now, relating this to the notion of complicated versus complex systems, it may be said that complex systems (CAS) are basically systems in which meta-levels have become active. In complicated systems the degree of interdependence is by definition low (Miller & Page, 2007). The degree of interdependence in turn could be low because no element has yet become a sufficient center of organization. In complicated systems it is likely that meta-levels are not active, even though there may be an incubation to make them more active. Or it could be the case that they were once active but have since become inactive because of having reached a plateau of stability. Technological dynamism creates high network connectedness versus technological stability that creates low network connectedness (Tatarynowicz et al., 2015). As meta-levels become active it can be seen that multiplicity, interdependence, and diversity naturally increase. Complicated systems could be viewed as an elementary stage in the formation of complex systems.

Considering the function Sig_E, summarized by Equation 7.1, this level of maturity may be linked to the emergence-matrix within Sig_E. Recalling Sig_E:

$$Sig_E = X \begin{bmatrix} C: Sig * mod\ (\int = 1) \\ F: Sig\ mod\ (c) \\ I: Sig\ mod\ (\int \overline{G, e, \pi}\ ) \\ M: Sig * mod\ (G) \\ V: Sig * mod\ (e) \\ P: Sig * mod\ (\pi) \end{bmatrix}$$

The Emergence-Matrix, as in Equation 14.2, can be thought of as:

$$\text{Emergence-Matrix} = \begin{bmatrix} C: Sig * mod\ (\int = 1) \\ F: Sig\ mod\ (c) \\ I: Sig\ mod\ (\int \overline{G, e, \pi}\ ) \\ M: Sig * mod\ (G) \\ V: Sig * mod\ (e) \\ P: Sig * mod\ (\pi) \end{bmatrix}$$

Eq 14.2: Emergence-Matrix

Keeping this emergence-matrix in mind, it may be said that a complicated system is one in which the signature is at a P, V, or M level. Hence, as depicted by Equation 14.3, where 'I' refers to the 'integral' level in equation 14.2:

$$Complicated_{System} = System\ (Sig_E: X < I)$$

Eq 14.3: Complicated System

By contrast a complex system, as in Equation 14.4, can be thought of as:

$$Complex_{System} = System\ (Sig_E: X \geq I)$$

Eq 14.4: Complex System

Chapter 15: General CAS Mathematical Operators

While there are several schools of thought relating to CAS, in general there is a core set of attributes considered to describe CAS. In their summary papers on CAS, Chan of MIT (Chan, 2001) and Dodder and Dare of MIT (Dodder & Dare, 2000) highlight the following representative aspects:

- Distributed control: no centralized single control mechanism that governs system behavior
- Connectivity: Complexity results from inter-relationship, inter-action, and inter-connectivity of elements within a system and between a system and its environment
- Co-evolution: Elements in a system can change based on their interactions with one another and with the environment
- Sensitive dependence on initial conditions: CAS are sensitive due to their dependence on initial conditions
- Emergent order: Complexity in CAS refers to the potential for emergent behavior in complex and unpredictable ways
- Far From Equilibrium: This phenomenon illustrates how systems that are forced to explore their space of possibilities will create different structures and new patterns of relationships
- State of paradox: CAS has dynamics of both chaos and order
- Multiple levels of organization: CAS tends to exist in many levels of organization in the sense that agents at one level are the building blocks for agents at the next level

- Unpredictability: CAS have a future that is hard to predict

As pointed out by Dodder & Dare the characteristics summarized above have themselves arisen from a study of similarities across CAS. In the natural world, brains, immune systems, ecologies, cells, developing embryos, and ant colonies, all fall under the category of CAS. In the human world, political parties, scientific communities and the economy, are examples of CAS. Human knowledge itself has been suggested to be an example of CAS (Thomas & Zaytseva, 2016), as has living cities (Gershenson, 2013). In his paper on 'Ecosystems and the Biosphere as Complex Adaptive Systems' (Levin, 1998) Levin points out that it is easy to find discussions of biospheres, ecosystems, economies, brains, organisms as CAS, and summarizes key characteristics as the sustained diversity and individuality of components; localized interactions among those components; an autonomous process that selects from among those components, based on the results of local interactions, a subset for replication or enhancement. In 'Economy as an evolving Complex System' (Arthur, 1997) Arthur states six properties true of all economies and all CAS to include: dispersed interaction, the absence of a global controller, cross-cutting hierarchical organization, continual adaptation, perpetual novelty, and far-from-equilibrium dynamics.

The afore-mentioned properties generally emanate from the prevalent bottom-up view of CAS.

Here 12 general properties or more precisely 'mathematical operators', arrived at deductively and thought to be true of CAS will be suggested. In contrast to the bottom-up and the rules-based views (Holland, 1992) these operators are the result of considering some of the dynamics of each of the levels and integration of the multiple levels of the mathematical model presented in this dissertation. These are non exhaustive, but rather are indicative of the nature of CAS.

Leveraging the quaternary basis of the mathematical model developed in the previous chapter these mathematical operators can be summarized in four categories that parallel the four-fold scheme: Presence, Power, Knowledge, and Nurturing. Since it is being suggested that any CAS is organized by such a quaternary system it perhaps also makes sense to consider the mathematical operators using such a lens.

Chapter 2 had suggested that there is an implicit order to Time and an implicit order to Space. Hence, any organization and the causality that it is subject to has to be reconsidered in light of the implicit time-space order that define the very fabric that animates all organization. If Space is plotted on the x-axis, and Time on the y-axis, and a general curve constructed characterized by a time-space boundary-n, as in Figure 15.1, then it can be assumed that the area, Time-Space Continuum, under the time-space boundary-n, is subject to the mathematics and properties being developed here.

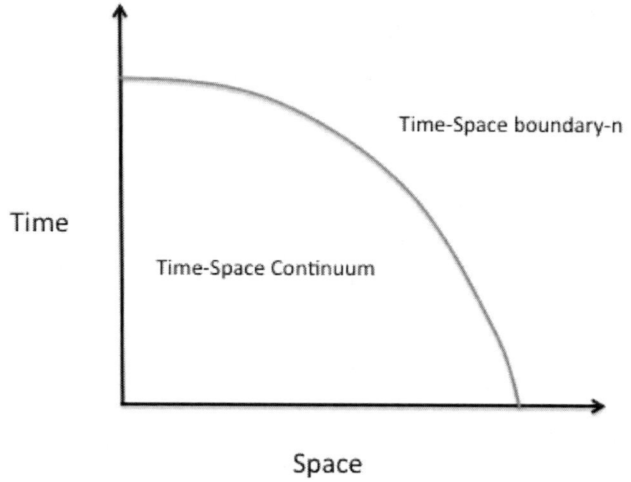

Figure 15.1 Time-Space Continuum

The following operators have to be considered in context to the General Equation for Innovation, Equation 9.6, derived in Chapter 9. This equation suggests that any CAS has implicit in it the urge to transform the untransformed layer, U, by opening to the influence of the meta-layers, M_1, M_2, and M_3. In so doing the very sources of innovation are altered and the visible characteristics of systems are transformed by the action of these sources of innovation. Several sets have already been suggested that explore these sources of innovation - $S_{System_{Pr}}$, S_{System_P}, S_{System_K}, and S_{System_N}, and the resultant characteristics of CAS - $Physical_T$, $Vital_T$, $Mental_T$, and $Integral_T$. The following sections leverage various aspects of the already derived mathematical model to frame sets of mathematical operators along each of the four basis of CAS that components or elements of CAS will be subject to in their journey toward transformation.

15.1 Presence-Based Mathematical Operators

Mathematical operators associated with presence would shed insight into the essential nature of any CAS element. Three representative presence-based mathematical operators include Fullness, Equality, and Uniqueness, and are described below. This presence-based set can be summarized by Equation 15.1.1:

$$Presence\_based_{Mathematical\_operators} \ni [Fullness, Equality, Uniqueness, \dots]$$

Eq 15.1.1: Set of Presence-based Mathematical Operators

Fullness refers to the possibility that every single point in the time-space continuum is informed by the four-fold fullness derived earlier in Chapter 3. Hence, in equation form it may be suggested that Fullness is the union (U) of $System_{Pr}$, $System_{P}$, $System_{K}$, and $System_{N}$. The full form of Fullness, as in Equation 15.1.2, would hence be:

$$Fullness = \bigcup \begin{bmatrix} \begin{bmatrix} \begin{matrix} Presence \\ \downarrow \\ Opportunity \end{matrix} \begin{bmatrix} P_L & \rightarrow & V_L \\ V_L & \rightarrow & M_L \end{bmatrix} \ \& \ Container_{\begin{bmatrix} System_P \\ System_K \\ System_N \end{bmatrix}} \end{bmatrix} \\ \begin{bmatrix} power > \sum_{TS=0}^{N} P_R * V_R * M_R \end{bmatrix} \\ \begin{bmatrix} \begin{matrix} Knowledge \\ \downarrow \\ Leverage \end{matrix} \begin{bmatrix} P_{I,C} \\ V_{I,C} \\ M_{I,C} \end{bmatrix} \rightarrow \begin{bmatrix} P_L & \rightarrow & V_L \\ V_L & \rightarrow & M_L \end{bmatrix} \end{bmatrix} \\ \left(\coprod_{Nurturing} \begin{pmatrix} P_- & M_+ \\ V_- & V_+ \\ M_- & P_+ \end{pmatrix} mod\,(r, \theta) \right) \end{bmatrix}$$

Eq 15.1.2: Full Form of Fullness

Equation 15.1.2 can be read from the top of the matrix. The first row highlights the dynamics of the system-presence, $System_{Pr}$, aspect of a point already derived in Chapter 3. The second row highlights the dynamics of the system-power, $System_P$, aspect of a point. The third row highlights the dynamics of the system-knowledge, $System_K$, aspect of a point. The fourth row highlights the dynamics of the system-nurturing, $System_N$, aspect of a point. The union, U, of all four rows signifies the fullness existing in every single point in any system.

A short-form of the equation for Fullness, as in Equation 15.1.3, would be:

$$Fullness = \bigcup \begin{bmatrix} System_{Pr} \\ System_P \\ System_K \\ System_N \end{bmatrix}$$

Eq 15.1.3: Short Form of Fullness

In general for any two points 'A' and 'B' it can be suggested that the Fullness behind A is the same as the Fullness behind B, as in Equation 15.1.4. This suggestion may also be arrived at by considering Einstein's General Theory of Relativity (Einstein, 1995) that states: "All bodies of reference K, K1 etc. are equivalent for the description of natural phenomena (formulation of the general laws of nature), whatever may be their state of motion." If we shrink these bodies of reference or coordinate systems to infinitesimal dimensions, thus approaching a 'point', this suggests that there is equivalence in that the general laws of nature are equally valid at any two points. Hence:

$$Fullness_A \equiv Fullness_B$$

Eq 15.1.4: Equivalence of Fullness

Equality refers to the possibility that since every point in a time-space continuum is always some expression of the underlying four-fold fullness, hence every point is equal to any other point. This implies that all expressions and developments share a fundamental equality with all other points. Depicting a point 'a' by $Point_A$ and a point 'b' by $Point_B$, then an equation for Equality, as in Equation 15.1.5, would be:

$$Equality\colon Point_A \equiv Point_B$$

Eq 15.1.5: Equality

Uniqueness refers to the possibility that any point is fundamentally unique. An easy way to envision this is to see that any point in a time-space continuum is a result of a unique time-space intersection. Hence, two points within a time-space continuum, A and B, can always be envisioned as having unique time and space coordinates:

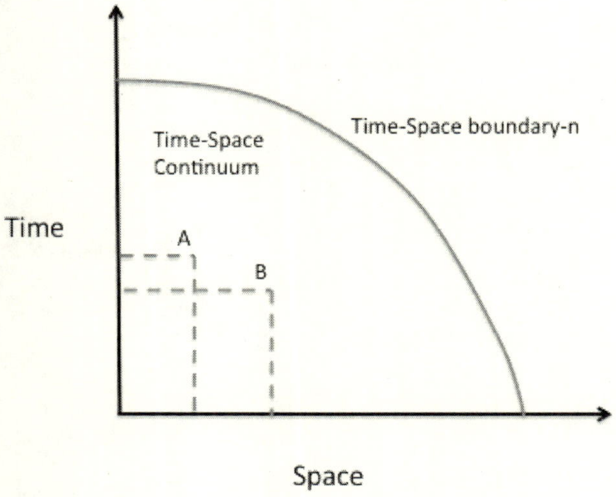

Figure 15.1.1 Uniqueness in Time-Space Continuum

But even for any organization, which can itself be considered a further development of a point or points through the action of time and space, the already proposed equation. Equation 6.5 derived in Chapter 6 that governs organizational signatures may be restated as an equation for uniqueness, as in Equation 15.1.6:

$$Uniqueness: Sig = \overline{Xa + Yb_{0-n}} \quad where \begin{bmatrix} X \in [S_{System_{P_r}}, S_{System_P}, S_{System_K}, S_{System_N}] \\ Y \in [S_{System_{P_r}}, S_{System_P}, S_{System_K}, S_{System_N}] \\ a, b \text{ are integers}; a > b \end{bmatrix}$$

Eq 15.1.6: Uniqueness

15.2 Power-Based Mathematical Operators

Power-based mathematical operators pertain to the fount of dynamism that will tend to determine an organization's practical action.

Mathematical operators related to 'power' give insight into how an organization will tend to meet circumstance and other organizations (Malik, 2009). Representative operators include Direction, Fractal, and Intersection and in contrast to Knowledge-based operators to be discussed in Section 15.3, tend to create the more visceral and immediate reactions to circumstance. Knowledge-based operators on the other hand tend to create action more in line with long-term or strategic plans. The integration between these different types of action and the notion of building such ambidexterity in organizations that combine incremental with perhaps more revolutionary shift is being recognized as increasingly important (Platzek et al, 2014). This power-based set can be summarized by Equation 15.2.1:

$$Power\_based_{Mathematical\_operators} \ni [Direction, Fractal, Intersection \ldots]$$

Eq 15.2.1: Set of Power-based Mathematical Operators

Direction refers to the possibility that direction at any possible system bifurcation point is not random and not determined either. Rather it can be thought of as a qualified determinism and is a function of applying DI_V and DI_H to the set of possibilities existing at a bifurcation point. As per the functioning of DI_V and DI_H discussed in detail in Chapter 11, it is the strongest or most 'powerful' possibility that will tend to determine what will emerge as an organization meets circumstance and other organizations. The equations have already been defined previously Chapter 11, and is adapted in Equation 15.2.2:

$$Direction: Org\_Dir = DI\left(\begin{bmatrix} M_3 \rightarrow System_X \\ (\uparrow F \rightarrow I) \\ M_2 \rightarrow S_{System_X} \\ (\uparrow Sig \rightarrow F) \\ M_1 \rightarrow Sig_x \\ (\uparrow > P_{P)} \\ U \rightarrow x_U \end{bmatrix}_{x=p,v,m,i}\right) \rightarrow$$

$$x\_matrix_{strongest} @ level_{strongest}$$

Eq 15.2.2: Direction

Fractal refers to the possibility that as organizational complexity increases, the base orientation, orientation-x, where x could be physical, vital, mental, or integral, of an average organization at some level of complexity 'n', will tend to determine the orientation of an organization at a level of complexity 'n+1'. Likewise the orientation or an organization at level of complexity 'n+1' will tend to determine the orientation of an organization at level of complexity 'n'. This is summarized by Equation 15.2.3:

$$Fractal: \ Orientation_x @ Complexity_n \leftrightarrow Orientation_x @ Complexity_{n+1}$$

Eq 15.2.3: Fractal

Organizational complexity refers to an order of magnitude change as in from a person to a team, of from a team to a business unit, and so on, for example. In Nature such fractal arrangements abound in the way the human body is constructed to the very structure of galaxies (Briggs, 1992). This notion has been suggested to exist in complex behavioral systems as well as described in some detail in books such as The Fractal Organization (Malik, 2015), The Fractal Organization (Hoverstadt, 2008), and The Misbehavior of Markets (Mandelbrot & Hudson, 2006). This notion relates to 'power' in that it is the patterns at one level that will tend to determine the patterns at another level, often preempting what may be a more logical choice based on reason. This kind of behavior has been suggested as causing cyclic fluctuations in stock and other markets where greed and fear often trump more intelligent and rational choices (Frost, 2005). Greed rises until fear sets in. Fear rises until greed sets in.

Intersection occurs when two organizations shift orientations to the next successive level due to the shock of interaction. Hence if an organization is at a physical orientation, it may be shifted to a vital orientation when intersection occurs, as in Equation 15.2.4, where the function 'Next Element' extracts the next element from the Set S comprising of the elements (physical, vital, mental, integral). Examples of such phenomenon abound where failure to make a shift results in shock of conflict repeating itself endlessly. Such a process with applicability at multiple levels of complexity has been captured by the series of books on crucial conversations (Patterson et al, 2011).

$$Intersection\text{: } Organization\ Orientation \\ \rightarrow Next\ Element\ (S)$$

Eq 15.2.4: Intersection

15.3 Knowledge-Based Mathematical Operators

Knowledge-based mathematical operators have to do with how organizations tend to develop by existing or toward increasing knowledge over time.

Representative knowledge properties discussed in this section include Alternative, Flowering, and Higher, which as suggested in Section 15.2 are of a different nature than 'power' or dynamism-based properties that tend to determine an organization's visceral or immediate reaction to the market place. This knowledge-based set can be summarized by Equation 15.3.1:

$$Knowledge\_based_{Mathematical\_operators} \ni [Alternative, Flowering, Higher \dots]$$

Eq 15.3.1: Set of Knowledge-based Mathematical Operators

Alternative refers to an alternative narrative that an organization will tend to embed itself in. These alternative narratives relate to the physical, the vital, the mental and the integral orientations introduced earlier and further elaborated by means of the derived equations in Chapter 9. These narratives can easily become fixed and can strongly influence the entire internal and external orientation of an organization. The book 'The Fractal Organization: The Future of Enterprise' suggests a theory of such narratives with their consequent effect on practical action (Malik, 2015). The alternative narratives are best described adapting the generalized equation derived for Innovation, Equation 9.6. Hence, Equation 15.3.2:

$$Alternative: Innovation_{orientation-x} = \begin{bmatrix} M_3 \rightarrow System_X \\ (\uparrow F \rightarrow I) \\ M_2 \rightarrow S_{System_X} \\ (\uparrow Sig \rightarrow F) \\ M_1 \rightarrow Sig_x \\ (\uparrow > P_x) \\ U \rightarrow x_U \end{bmatrix} TC \rightarrow x_T, where \begin{bmatrix} x_U \ni [\dots] \\ x_T \ni [\dots] \end{bmatrix}$$

Eq 15.3.2: Alternative

Where, 'x' can be thought of as an element from the Set of Orientations:

$$x \in (physical, vital, mental, integral)$$

'X' can be thought of as an element from the Set of System-level architectural forces. Hence:

$$X \in (Presence, Power, Knowledge, Nurturing)$$

Flowering refers to the possibility that any time-space boundary-n, depicted as TS_n will have more potential or possibility associated with it than a time-space boundary-(n-1), depicted as TS_{n-1}. Putting this into equation format as in Equation 15.3.3:

$$Flowering: Possibility_{TS_n} > Possibility_{TS_{n-1}}$$

Eq 15.3.3: Flowering

Higher refers to the possibility that over time the direction will always tend to move to a higher meta-level. This can be summarized by using the notion of the core-matrix summarized earlier in Equation 9.7. Adapting this hence, as Equation 15.3.4:

$$Higher: Upward \begin{bmatrix} M_3 \rightarrow System_X \\ (\uparrow F \rightarrow I) \\ M_2 \rightarrow S_{System_X} \\ (\uparrow Sig \rightarrow F) \\ M_1 \rightarrow Sig_x \\ (\uparrow > P_{x)} \\ U \rightarrow x_U \end{bmatrix}$$

Eq 15.3.4: Higher

Note also that while many organizations are practically at the untransformed level for a long time there is still movement within that level that can generally be depicted as a change from a predominantly physical-orientation to a more mental-orientation as was discussed in Chapter 2. In general this shift in orientation is implied as more and more fixed patterns are overcome: $\uparrow > P_x$, where P_x refers to patterns along an orientation 'x' where x is an element from the set: (physical, vital, mental, integral).

15.4 Nurturing-Based Mathematical Operators

Nurturing-based mathematical operators have to do with the nature of relationship within CAS. These relationships are posited as being of a nurturing nature and emanate from the notion of 'nurturing' as an organizing class for CAS.

Representative mathematical-operators include Remember, Linking, and Relate. This nurturing-based set can be summarized by Equation 15.4.1:

$$Nurturing\_based_{Mathematical\_operators} \ni [Remember, Linking, Relate \ldots]$$

Eq 15.4.1: Set of Nurturing-based Mathematical Operators

Remember has to do with remembering that there is something in each organization that existed before the existence of any organization, and further, that there is something in each organization that exists in every other organization. This can be thought of going back to a time-space moment of zero, and subsequently of expanding into the Time-Space Continuum keeping that connection in mind. There is something in each organization that exists in every organization and highlights a special way to relate to the underlying system:

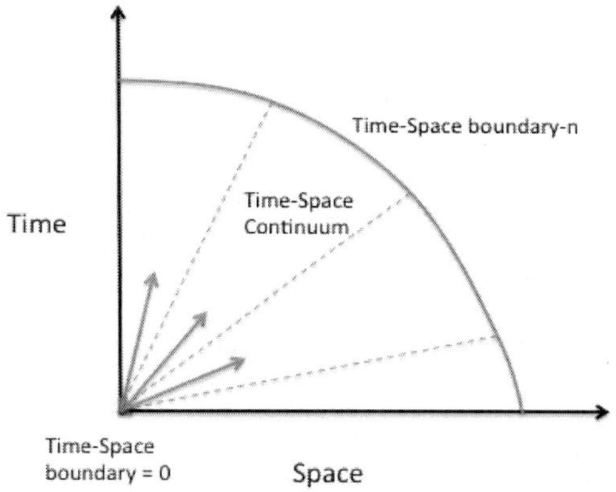

Figure 15.4.1 Remembrance in Time-Space Continuum

This may be depicted by Equation 15.4.2:

$$Remember: Ubiquity \begin{bmatrix} \xleftarrow[TS=0]{} \\ \xrightarrow[TS>0]{} \end{bmatrix}$$

Eq 15.4.2: Remember

Where the condition of going back (←) before any organization existed, TS = 0, is invoked as all developments proceed (→), TS > 0, to create a sense of ubiquity. The sense of ubiquity is a remembrance.

This notion may be akin to the concept that everything that is in the universe emanated from the Big Bang and that we are all recycled stardust (Swimme, 2001).

Link refers to the condition whereby in any time-space coordinate, irrespective of the level of untransformed reality (U), any present state can be consciously linked to the underlying ubiquitous system. This can be depicted by Equation 15.4.3 where the conditions that usually need to be in place for a meta-layer to actively influence layer U disappear. There may be an attitude or receptiveness on the part of the element at U that allows such linking to take place. Hence:

$$Link: \begin{bmatrix} M_3 & \rightarrow & System_X \\ & \uparrow\downarrow & \\ M_2 & \rightarrow & S_{System_X} \\ & \uparrow\downarrow & \\ M_1 & \rightarrow & Sig_x \\ & \uparrow\downarrow & \\ U & \rightarrow & x_U \end{bmatrix}$$

Eq 15.4.3: Linking

Relate is a way to relate to the System so as to offer or surrender activities any kind of organization is involved in, to the System, and hence the Fullness or intelligence embedded in every point. This can be depicted by an Offer function, such that the first or relatively untransformed element in the function, depicted by x_U, is being offered to the second one, the union of the four-fold intelligence embedded in each point and depicted by the union function U[ ], as in Equation 15.4.4:

$$Relate: Offer\left(x_U, \bigcup \begin{matrix} System_{Pr} \\ System_P \\ System_K \\ System_N \end{matrix} \right)$$

Eq 15.4.4: Relate

REFERENCES

1. Aral, M.M., 2016. Transition from Simple, Complicated to Complex Systems.*Transition*.
2. Arkani-Hamed, Nima; Bourjaily, Jacob L.; Cachazo, Freddy; Goncharov, Alexander B.; Postnikov, Alexander; Trnka, Jaroslav (2012). "Scattering Amplitudes and the Positive Grassmannian". arXiv:1212.5605
3. Arthur, B, Durlauf N, Lane D. 1997. The economy as an evolving complex system II. Addison-Wesley. Reading (MA)
4. Bar-Yam, Y. 2016. From big data to important information. Complexity, 21: 73–98. doi:10.1002/cplx.21785
5. Briggs, J. 1992. Fractals: The Patterns of Chaos. Simon & Schuster: New York.
6. Chan, Serena. 2001. Complex Adaptive Systems. MIT. ESD-83 Research Seminar in Engineering Systems
7. De Broglie, L. 1929. The Wave Nature of the Electron. Nobel Lecture. http://www.nobelprize.org/nobel_prizes/physics/laureates/1929/broglie-lecture.pdf
8. Deep Order Mathematics Videos. 2016. Deep Order Technologies. http://www.deepordertechnologies.com/new-index
9. Dodder, R & Dare, R. 2000. Complex Adaptive Systems and Complexity Theory: Inter-related Knowledge Domains. ESD.83: Research Seminar in Engineering Systems. MIT
10. Efatmaneshnik, M, Nilchiani, R, Heydari, B. 2012. "From complicated to complex uncertainties in system of systems," *Systems Conference (SysCon), 2012 IEEE International*, Vancouver, BC, 2012, pp. 1-6.
11. Einstein, A. 1995. Relativity: The Special and General Theory. New York: Broadway Books.
12. Ekspong. 2014. "The Dual Nature of Light as Reflected in the Nobel Archives". *Nobelprize.org.* Nobel Media AB 2014. Web. 15 Oct 2016. http://www.nobelprize.org/nobel_prizes/the

mes/physics/ekspong/

13. Feynman, RP. 1985. QED The Strange Theory of Light and Matter. New Jersey: Princeton University Press
14. Flack, J.C. 2014. Life's Information Hierarchy. Santa Fe Institute Bulletin. April 2014.
15. Frenkel, E. 2013. Love & Math: The Heart of Hidden Reality. Basic Books: New York
16. Frost, R. 2005. Elliott Wave Principle: Key to Market Behavior. New Classics Library: Georgia
17. Gershenson, C. 2013. Living in Living Cities. Artificial Life 19:3_4, 401-420
18. Hawking, Stephen. 1988. A Brief History of Time. New York: Bantam Books
19. Holland, John. 1992. Complex Adaptive Systems. Daedalus, Vol. 121, No. 1, A New Era in Computation (Winter, 1992), pp. 17-30 Published by: The MIT Press on behalf of American Academy of Arts & Sciences
20. Hoverstadt, P. 2008. The Fractal Organization: Creating Sustainable Organizations with the Viable System Model. Chichester: John Wiley & Sons
21. Ismail, S. 2014. Exponential Organizations: Why new organizations are ten times better, faster, and cheaper than yours (and what to do about it). Diversion Books: New York
22. Jeans, J. 1932. The Mysterious Universe. Cambridge University Press.
23. Johnson, S. 2010. Where Good Ideas Come From: The Natural History of Innovation. Riverhead Books: New York
24. Kaufmann, S. 1995. At Home in the Universe. New York: Oxford University Press.
25. Kurzweil, R. 2013. How to Create a Mind: The Secret of Human Thought Revealed. New York: Penguin Books
26. Laszlo, E. 2014. The Self-Actualizing Cosmos: The Akasha Revolution in Science and Human Consciousness. Inner Traditions: Vermont
27. Levin, S. 1998. Ecosystems and the Biosphere as Complex Adaptive Systems. Ecosystems. Springer-Verlag

28. Lorentz, H.A. 1925. The Science of Nature. Vol. 25, p 1008. Springer
29. Malik, P. 2003. Business Transformation Through the Creation of a Complex Adaptive System. Journal of Human Values 9:2. New Delhi: Sage Publications
30. Malik, P. 2004. Introduction to Fractal Dynamics. Journal of Human Values 10:2. New Delhi: Sage Publications
31. Malik, P. 2009. Connecting Inner Power with Global Change: The Fractal Ladder. New Delhi: Sage Publications
32. Malik, P. 2011. Redesigning the Stock Market: A Fractal Approach. New Delhi: Sage Publications
33. Malik, P. 2013. Current Economic Challenges: Making Sense of Market Crises. Integral Review. 9(3), 50-63
34. Malik, P. 2014. The Flower Chronicles: A Radical Approach to Systems and Organizational Development. New Delhi: Aurosoorya
35. Malik, P. 2015. The Fractal Organization. New Delhi: Sage
36. Mandelbrot, B, Hudson, R. 2006. The Misbehavior of Markets: A Fractal View of Risk, Ruin, and Reward. New York: Basic Books
37. Meadows, D. 2008. Thinking in Systems. White River Junction: Chelsea Green Publishing.
38. Miller, J.H, Page, S. E., 2007. Complex Adaptive Systems: An Introduction to Computational Models of Social Life. Princeton University Press. New Jersey.
39. Mora, C, Tittensor, D, Adl, S, Simpson, A, Worm, B. 2011. How Many Species Are There on Earth and in the Ocean? Plos.org. http://journals.plos.org/plosbiology/article?id=10.1371/journal.pbio.1001127#abstract0
40. Ogbonna, E, Harris, L. 2000. Leadership Style, Organizational Culture and Performance: Empirical Evidence from UK Companies. International Journal of Human Resources. Taylor & Francis: London
41. Ogburn, W. Tomas, D. 1922. Are Inventions Inevitable? A Note on Social Evolution. Political Science Quarterly 37, no.1

42. Patterson, K, Grenny, J. Crucial Conversations: Tools for Talking when Stakes are High. 2011. McGraw-Hill: New York.

43. Pauli, W. 1964. Nobel Lectures, Physics 1942 – 1962. Elsevier Publishing Company. Amsterdam.

44. Pearson, R. 1997. "Frontier Perspectives", Journal of the Center for Frontier Sciences at Temple University. Spring/Summer 1997, Volume 6, Number 2, ISBN: 1062-4767.

45. Perkowitz, S. 2011. Slow Light. London: Imperial College Press

46. Planck, M. 1933. Where is Science Going? Ox Bow Press. Connecticut.

47. Platzek, B., Pretorius L., & Winzker, D, 2014, The Vital Entrepreneurial Learning Organization: A Corporate Mindset for Entrepreneurial Change Management, International Journal of Innovation and Technology Management, Vol. 11, No. 6 (2014) 1450044, World Scientific Publishing Company

48. Portugali, J., 2012. *Self-organization and the city*. New York: Springer Science & Business Media.

49. Prigogine, Ilya. 1967. Introduction to Thermodynamics of Irreversible Processes. J. Wiley

50. Prigogine, I. 1977. Time, Structure, and Fluctuations. *Nobelprize.org.* Nobel Media AB 2014. Web. 5 Mar 2016. <http://www.nobelprize.org/nobel_prizes/chemistry/laureates/1977/prigogine-lecture.html>

51. Reeves, M, Zeng, M, Venjara, A. 2015. The Self-Tuning Enterprise. Harvard Business Review June 2015. Cambridge: Harvard University Press

52. Salkind, N. 2007. Encyclopedia of Measurement and Statistics. Thousand Oaks: Sage Publications.

53. Sargut, G, McGrath, R. Learning to Live with Complexity. Harvard Business review. September 2011.

54. Seligman, M. 2011. Flourish: A Visionary New Understanding of Happiness and Well-Being. Free Press: New York

55. Snyder, M. 2010. Stanford Medicine.

http://med.stanford.edu/news/all-news/2010/03/what-makes-you-unique-not-genes-so-much-as-surrounding-sequences-says-stanford-study.html#.html

56. Sri Aurobindo. 1971. Social and Political Thought. Sri Aurobindo Ashram Press: Pondicherry
57. Sternberg, R. 2012. The Assessment of Creativity: An Investment-Based Approach. Creativity Research Journal. Taylor & Francis: London
58. Swimme, B. 2001. The Universe is a Green Dragon. Bear & Company: Rochester
59. Tatarynowicz, A, Sytch, M, Gulati, R. 2015. Environmental Demands and the Emergence of Social Structure. Administrative Science Quarterly. Vol 61, Issue 1, pp. 52 – 86. First published date: September-25-2015. 10.1177/0001839215609083
60. Thomas, D, Brown, J. 2009. Why Virtual Worlds Can Matter. International Journal of Learning and Media 2009 1:1, 37-49
61. Thomas, J, Zaytseva, A. 2016. Mapping complexity/Human knowledge as a complex adaptive system. Complexity, 21: 207–234. doi:10.1002/cplx.21799
62. Turing, A. 1952. The Chemical Basis of Morphogenesis. Philosophical Transactions of the Royal Society of London B: Biological Sciences
63. Van Wylen, G.J, Sonntag, R.E (1985), Fundamentals of Classical Thermodynamics, Section 5.5 (3rd edition), John Wiley & Sons Inc. New York.
64. Weinberg, S. 1994. Dreams of a Final Theory: The Scientist's Search for the Ultimate Laws of Nature. Vintage Press: New York
65. Whitaker, A._2006. Einstein, Bohr and the Quantum Dilemma: From Quantum Theory to Quantum Information. Cambridge: Cambridge University Press
66. Wilczek, F. 2016. A Beautiful Question: Finding Nature's Deep Design. New York: Penguin Books
67. Willems, J. 1998. State & Storage. Perspectives

in Control: Theory & Applications (ed. Dorothee Normand-Cyrot). Springer: New York.

68. Yates, F.E. 2012. *Self-organizing systems: The emergence of order.* New York: Springer Science & Business Media.

SPECIAL THANKS

Special Thanks to Professor Dr. Leon Pretorius for leading me to a higher level of academic rigor through his Socratic approach during my doctoral work.

Made in United States
North Haven, CT
06 November 2026

21880160R00060